MILESTONES OF
SCIENCE AND TECHNOLOGY

Driving wheel of the steam locomotive Rocket
(*See* STEPHENSON'S *Rocket*)

MILESTONES OF SCIENCE AND TECHNOLOGY

MAKING THE MODERN WORLD

REVISED AND EXPANDED SECOND EDITION

PETER MORRIS, Executive Editor

JAMES C. HART, Managing Editor

LESLEY HENDERSON, Further Reading Editor

PHILIP SAYER, Photographs

KWS
PUBLISHERS
CHICAGO ✳ LONDON

In association with the

SCIENCE MUSEUM
LONDON

Published by:

KWS Publishers

1516 North State Parkway

Chicago, Illinois 60610, U.S.A.

This Second Revised and Expanded edition first published in 2013

First edition published in 1992 by John Murray (Publishers) Ltd.

A catalogue record for this book is available from the Library of Congress
and from the British Library

ISBN 978-0-9817736-5-0

On the front cover: Sir William Herschel's seven-foot telescope, c. 1781

On the back cover: Panhard et Levassor Car, 1895

Typeset in Goudy Oldstyle and ITC Stone Sans

Printed and bound in the USA by CJK Printing

Book design and production by Studio 31, Inc.

CONTENTS

*"Sketch taken from Whitworth's stand of machine tools,
for planning, slotting, drilling, boring, etc…"
at the Great Exhibition, 1851*

LONDON'S SCIENCE MUSEUM:
AN INTRODUCTION

ROBERT BUD

Science is not just for scientists. A founding curator of London's Science Museum wrote in 1926 that the objectives of the Museum's exhibitions were to persuade the visitor that science should be close to "his business and his bosom." The continuing buzz, chatter and comments of the thousands of people of all ages who have animated the Museum's galleries seven days a week, month after month, for a hundred years, have attested to the attractiveness of this mission. In 1928, the year in which the new building opened, a million visitors came to the Museum for the first time. Eighty years later, two to three million visitors a year has been the norm. In its own lifetime the Museum has played a significant role in that two-century-long period in which scientific and industrial society, beginning in Europe—and in the late 18th century, Great Britain in particular—has become a global phenomenon that has transformed the world.

Despite the long and noble ancestry of museums of all kinds, it was not before the 1870's that there was any kind of action in Britain to found the world's first museum specifically devoted to science. France had long had a museum of technology (arts et métiers), and many universities conserved with devotion old collections accumulated by revered professors. But what happened in London was more than a jealous reaction to the French or organized sentiment for lost colleagues. Germany and the United States had challenged Great Britain's industrial supremacy, and Germany had already long been a center of pilgrimage for scientists. Japan was coming out of isolation. Global markets for the rapidly developing products of industry were being hotly contested. With globalization had come a growing awareness of the world-historical importance, in Britain a century earlier, of such inventions as the steam engine, automated textile production and iron technology. Well-known late 19th century scholars, teachers and publicists, such as Thomas Huxley and Lyon Playfair, attributed these inventions to the wise application of science.

Queen Victoria's consort, Prince Albert, had championed a museum interpreting the route from

HRH Prince Albert (1819–61)

past to future glory by displaying a combination of manufactures with the arts. But Albert died suddenly in 1861, and in the following decade his approach and philosophy fell out of favor. The so called South Kensington Museum, combining a miscellaneous combination of collections relating to the applied arts, certain manufactures and education, which had been founded under Albert's aegis in 1857, was found to be rich in parts but not fit for what many people felt the purpose of a science museum ought to be. Then, its founding director, Henry Cole, retired in 1873. Within two years, yet another of the pillars of the old South Kensington establishment passed from the scene: Bennet Woodcroft—champion of the neighboring Patent Office Museum, which had assembled examples of Great Britain's manufacturing achievements—also retired. Without a dynamic collector at its helm,

the Patent Office wished to divest itself of its inherited great collection.

By the mid-1870's, in other words, there was a growing agreement that the old model for a science museum really hadn't worked. A science museum, to be relevant to the late 19th century, had to tell stories of past scientific achievements, proclaim the potential of current scientific practices, and celebrate practical applications of those practices. The exhibitions should inspire the country as a whole to cope with dimly imagined trials to come, inform a new generation of scientifically literate people, and promote the standing of scientists within modern society. Science, as such, was to be broadly interpreted, so that the development of the steam engine could be seen as the application of science. This example was particularly relevant because of the economic importance of steam and Britain's unchallenged right to be regarded as the pioneer of steam power. How such an invention happened in detail was less important than the power of the message, which was to be based on its success—not only the invention of the steam engine but also the success of other inventions that involved a practical application of scientific principles.

In 1874 the Royal Commission on Scientific Instruction and the Advancement of Science produced its final report on the concept of a new science museum. The secretary of the commission, Norman Lockyer, who wrote the report, was an energetic civil servant, an amateur astronomer and the pioneering editor of the new scientific journal *Nature*. He included an appeal for better recognition of science: "We accordingly recommend the formation of a collection of physical and mechanical instruments; and we submit for consideration whether it may not be expedient that this collection, the collection of the Patent Office Museum, and that of the Science and Education Department of the South Kensington Museum should be united and placed under the authority of a Minister of State."

Lockyer was asked to lead the next stage of development. A committee headed by the Lord Chancellor, and including the presidents of professional institutions and learned societies, was formed to advise him. The collection was to include "not only apparatus for teaching and investigation, but also such as possessed historic interest on account of the persons by whom, or the researches in which, it had been employed." And the combined collection's first exhibition, opened in May 1876 by Queen Victoria, was to be international in scope. The great commitment of other European countries, including France and the newly-united Germany, was signaled by the translation of the thousand-page catalogue into both French and German. The exhibition was held in existing buildings across the Exhibition Road from the South Kensington and Patent Office museums that were on its eastern side. The so called loan exhibition, which was curated by Lockyer, consisted of 20,000 objects, which is more than the total number of artifacts exhibited in the Science Museum today. It included both historical and period examples of science and of its "applications." James Watt's steam engine and Stephenson's *Rocket* locomotive from the Patent Office Museum were examples of "application."

Once assembled on the western side of the Exhibition Road, the displays of scientific practice and application stayed where they were. In 1884 the Patent Office Museum collections were formally transferred to the Science and Art Department of the South Kensington Museum. The Royal Commission for the Great Exhibition of 1851, which owned the land on both sides of Exhibition Road, offered £100,000 for the building of a "science museum." This vision was not achieved without a struggle, however. A letter to *The Times* newspaper in January 1877 warned, "What might not a 'Science Museum' be expected to contain? A huge and incongruous agglomeration of objects of the greatest diversity from the four quarters of the globe, which it would require a well-paid army of curators to arrange and keep in order."

This was not an isolated criticism; in its own way, it was prophetic. The 1880's and 1890's involved a relentless stream of committees and commissions, the purpose of which was to examine

Bennet Woodcroft (1803–79), founder of the Patent Office Museum

the purpose, the future, and the building needs of a separate science museum. It was during this period that the same people who were calling for the establishment of a science museum were also advocating state funding of science—what was called "the endowment of research" movement. Skeptics, who were numerous, lampooned the enterprise, satirizing it as "the research of endowment," and their derision was also extended to the idea of a science museum. Nevertheless, the name Science Museum was widely used from the early 1880's, and in 1893 separate directors were named for the Arts and Science collections, though they still shared an administrative heart and an identity as parts of the South Kensington Museum. This museum was renamed the Victoria and Albert Museum in 1899 when Queen Victoria laid the foundation stone for a new building on the east side of the Exhibition Road.

With changes in the name came changes in method. Widespread dissatisfaction with the quality of the arts acquisitions led to the appointment of expert curators (as at the British Museum) rather than merely collection administrators. This change was implemented for both divisions of the museum, but the tension between a heritage-oriented art division and a future-oriented science division was becoming too intense for the two to remain components within a single organization. Supporters of a separate Science Museum now included the powerful education civil servant Robert Morant, as well as his boss, Lord Runciman, the Secretary of the Board of Education, whose family had long been leading shipbuilders in the northeast of England. As well, Lockyer, the long-serving and combative editor of *Nature*, was still fighting hard for a separate museum, and he was supported by a fast-growing scientific community. The opening of the Victoria and Albert Museum in June 1909 provided the opportunity for independence. Morant "recalled" that the late Queen had wanted the name Victoria and Albert Museum to apply *only* to the building now dedicated to the arts. He maintained that the use of the term to cover the science collections as well had been a decade-long mistake. All of this agitation in high places had the desired effect: June 26, 1909, the day of the V&A's opening, was also the birth date of an entirely independent Science Museum.

The British government appointed a committee under the chairmanship of Sir Hugh Bell, a man prominent in the iron and steel industry (and, incidentally, father-in-law to Runciman's deputy Charles Trevelyan), to consider what ought to be the aims and goals of the Science Museum and how they might best be achieved. The Bell Committee completed its report in 1912: its recommendations were accepted. A plot between Exhibition Road on the east and Queens Gate on the west would gradually be filled with new buildings that together would constitute the Science Museum. Work started on the East Block of a new building the following year. Indeed, the Bell Report is the blueprint for the foundation of the Science Museum and for its development up to the present day. Its recommendations—which still stand as guiding principles for the Science Museum and its development—have been only in part implemented. Even now, despite the considerable scale and diversity of the Science Museum and its activities, and the tremendous growth in its collections, only two-thirds of what was planned eighty years ago has yet been built.

In other words: in the years before World War I, plans were being made that would shape the institution for the next century—and these plans were based on visions that could be traced back to mid-Victorian England. The continuity between the two eras was maintained, pre-eminently, by the museum's brilliant director, Sir Henry Lyons. Then a royal engineer in the British Army, Lyons had been sent to Egypt in 1890 to help the British-backed government there. Soon after arriving, he was given a surprising job: he was seconded to work with a visiting scientist, Norman Lockyer. Thus, a close association was forged with Lockyer, who was then exploring his hypothesis that the temples of the Aswan region were oriented astronomically. For two years Lockyer and Lyons worked closely together, their work resulting in Lockyer's carefully researched (though subsequently discredited) 1894 volume, *The Dawn of Astronomy: A Study of the Temple-Worship and Mythology of the Ancient Egyptians*.

Lyons then returned to his government work as a surveyor of the Egyptian desert. He had, however, now become very interested in archaeology, and, with Lockyer's help, he was elected to the Royal Society in 1906. He returned to Britain in 1909. In 1911, Francis Grant Ogilvie, the newly appointed director of the Science Museum, requested an assistant who would not only be the Secretary to the Museum's new advisory council but also, effectively, a deputy director. Lyons was chosen for this post. A decade later, in 1920, he succeeded Ogilvie as Director of the Museum at the same time that it was moving into its new building. Although the shell of the new building on Exhibition Road had

Entrance to the Patent Office Museum, 1863

been completed before World War I, the building was not completed and opened, under Lyons' leadership, until 1928. The opening by King George V of the Science Museum was the triumph of a campaign that had lasted half a century.

But much had changed since mid-Victorian times. Ironically, at the moment of success, the Museum's actual opening, new voices were already challenging this latest model of what a science museum ought to be. The support of industry was needed to fund and support exhibitions. Moreover, industrial research and technology had come to the fore during the war. Recently formed conglomerates based on new technologies—such as Imperial Chemical Industries and the General Electric Company (GEC)—supported research, but were more concerned with products coming to market than with the branding of British industry. Their support of a science museum did not come free. In 1923, the Science Museum acquired a "subtitle:" The National Museum of Science and Industry. The report of the 1929–30 Royal Commission on

Museums and Galleries urged the Science Museum to move more closely to industry. The new subtitle did not in fact become widely known; however, the Science Museum was driven by the tension created by the duality of its actions: teaching the principles of science at the same time that it was demonstrating the products of industry.

The Science Museum's early years were characterized by trying for a kind of reconciliation of those goals.

A host of temporary exhibitions were mounted. In 1931, a Children's Gallery was opened: it demonstrated the principles of science. It attempted to engage children in a way that one would now call "interactive." Its artifacts, ranging from pulleys in various arrangements that tested muscles, to a self-opening door, were designed to give physical expression to fundamental principles. In replying to a question from the Brooklyn Children's Museum as to the curriculum in place in the Children's Gallery, the Director replied, "None, thank God. The children are allowed to wonder as they will."

The 1933 Plastic Materials Exhibition, which demonstrated the uses of these new materials, attracted more than a million visitors. The tension between demonstrating science and exhibiting its industrial consequences was in the end a creative one, and it set the new Museum on a productive course. Sir Henry Lyons retired in 1933: his regime was acknowledged to have been outstanding. The institution had recently been described as one of the five most progressive museums in the Empire—possibly, first of the five. In just twelve years the number of annual visitors had increased from 400,000 in 1921 to 1,250,000 at Sir Henry's retirement. In his address to Sir Henry, Sir Richard Blazebrook, the Chairman of the Science Museum, reflected that Sir Henry had made the Museum a "treasure house of past achievements and an inspiring guide to future progress."

In endeavoring to preserve a balance between past and future, the Science Museum developed a distinctive collections policy. It acquired samples of the most recent scientific and technological achievement, such as the first radio tube/valve or the original Wright flyer; as well, it collected historical artifacts documenting the evolution of science. Missing links in the Ascent of Man (particularly Englishmen) were punctiliously filled in, as was every step in the development, say, of steam navigation or the measurement of electric current.

The Science Museum closed shortly after the outbreak of World War II. Its holdings were evacuated to safe places, and many of its staff became involved in some sort of war service. Only the library remained open, to satisfy wartime demand for scientific literature. In February 1946, after the war was over, the Science Museum reopened. Plans were now made to erect the second stage of the building envisioned by Bell, and by 1951 the Centre Block was partially complete, just in time to accommodate the Science Section of the Festival of Britain. The Centre Block was not finally completed, however, until 1961, ten years later.

Then began a period of twenty years of steady expansion. Further space was added in 1975. In 1980–81, objects from the Wellcome Museum of the History of Medicine, which the Wellcome Trust had placed with the Museum in 1976, were put on display. This outstanding collection, relating to medical history, a subject new to the Museum, was one of the most important acquisitions in its history, and it significantly broadened the portfolio of sciences that the Museum regarded itself as covering. The Wellcome Collection is strongly international: it reinforced the tendency of the Museum to consider scientific achievement throughout the world.

At the turn of the Millennium, the vision of the Bell Committee was given further affirmation with the opening, first, of a new wing, also partly funded by the Wellcome Trust and named the Wellcome Wing, which involved a huge westward expansion of the current building, and, second, the creation of the Dana Centre within a completely new building, the Wellcome-Wolfson Building, which reaches to the western edge of the original plot of ground dedicated to the Museum. Both of these new structures are dedicated to the interpretation of modern science. In the 21st century, physical exhibits have been complemented by online interpretations of the collections. The new Wellcome Wing was linked to the older galleries by a major exhibit in the space first occupied by the exhibition for the Festival of Britain. This exhibit—Making the Modern World—of which this book is a celebration—encapsulated the message of the Museum by presenting objects of scientific and technical innovation primarily from the 18th century industrial revolution to the 21st century information and biotechnological eras. This book, *Milestones of Science and Technology*, includes essays and photographs of 112 of these objects.

The Library, too, was given a new direction during the later 20th century. It had lost its national lending function during the 1960's. Meanwhile, its neighbor, Imperial College, had considerably developed its own libraries. In 1971, the government directive that led to the establishment of the British Library also mandated that the Science Museum Library should be developed as a national reference library of the history of science and technology, with particular emphasis on supporting the work of the curators in the other departments of the Museum. The Science Museum Library moved into new accommodations that are contiguous with the Lyon Playfair Library of Imperial College, and there is now a close working relationship between the two.

There were still other changes to come for the Library. Because of an increasing demand for space in South Kensington, a strategic review, conducted during 2002–05, recommended the re-location of the Library's original print collections and archives from the library's stacks and other storage areas in the Museum to the Museum's airfield site, in Wroughton, near Swindon, Wiltshire. Here they joined many of the large objects in the Museum's collections, which were already stored there in several World War II aircraft hangars. During 2006, a large archive and library storage space was built inside one of these hangars, and some ninety percent of the collection moved from the poor-quality London storage areas to this new high-quality accommodation, nearly filling its eighteen kilometers of stacks. During 2007, an adjacent building was completely refurbished to form a new library/archives reading room, with study spaces for thirty researchers, staff facilities, and a further eight kilometers of good-quality storage space. The collection in South Kensington now serves as a contextual introduction to the collections at Wroughton, as well as itself being of international significance.

For most of its history, the Museum's activities were concentrated within its building in South Kensington, even though substantial reserve collections were housed in warehouses elsewhere in London. In the early 1970's, however, a significant development took place: under the terms of the Transport Act 1968, the collections of railway material that had been accumulated since 1951 under the auspices of the British Transport Commission were transferred to the Department of Education and Science, of which the Science Museum

The Crystal Palace, completed in 1851 to house the Great Exhibition of the Works of Industry of all Nations

was by then a part. It was decided that a new national museum should be developed to house this outstanding railway material, and after a long debate about where its location ought to be, it was decided to convert the North Motive Power Depot at York, which had recently closed as a result of the changeover from steam to diesel-electric traction, to this purpose. York was a good choice for a new National Railway Museum. A railway center in its own right, with a long and distinguished history as a rail hub, it was also home to Great Britain's first railway museum, opened in 1928 by the London and North Eastern Railway in small premises on Queen Street. With this decision to establish the National Railway Museum at York, under the direction of the Science Museum, the National Railway Collection had a new and permanent home. The Queen Street museum was closed, and in September 1974 the National Railway Museum in York was opened by the Duke of Edinburgh.

A decade later, the Science Museum expanded in Yorkshire once again, this time creating a new National Museum of Photography, Film and Television in Bradford, the first section of which opened in 1983. It marks another new emphasis for the Museum—on the post-industrial category of information. This museum now houses most of the Science Museum's photographic material, including the outstanding collection of Talbot images, transferred from South Kensington in 1989. In 2007 this museum was renamed: it is now known as the National Media Museum.

While the Museum expanded, the country was also changing around it. The new Conservative administration elected in 1979 radically rethought the role of the State. The Science Museum had been formed as part of the Board of Education, and until 1983 it was part of that board's administrative successor within the U.K. Civil Service, the Department of Education and Science. At the end of 1983, however, under the terms of the National Heritage Act, the Museum ceased to be part of a government department and its administration was transferred to a board of trustees, which now had the responsibility for "balancing the books." The idea of the Science Museum, or any museum, as an agent of government policy, funded to implement national goals, was clearly abandoned. Now, though the government did indeed still contribute a subsidy, the institution was expected to fund its own needs. The Science Museum was forced to begin charging entrance fees in 1987, and this practice lasted a decade, until the next Labour administration again subsidized free admission for the public.

Both the National Railway Museum and the National Media Museum had been founded with a status much less grand than their titles. Both were "outstations" of the Science Museum; they could perhaps be compared to provincial colonies. At the beginning of the 21st century, however, that status was radically changed. The old title—National Museum of Science and Industry—was now redeployed to describe a commonwealth of three equals. The three equals—the Science Museum, the National Railway Museum, and the National Media Museum—now serve an audience of some five million people each year.

The activities and the collections of the Science Museum at the beginning of the 21st century are inspired by the concerns of the society in which it exists and by its ambitions for the future: these goals are reflected in the Museum's recent activities. Yet, at the same time, there have been remarkable continuities, a desire not to leave the past behind. Ten objects were chosen by curators as icons for the centenary year, 2009–10. They include: Stephenson's *Rocket*, acquired by Bennet Woodcroft for the Patent Office Museum and absorbed into the new Science Museum in 1885, as well as a Newcomen engine acquired in 1920. The other great icon of the 19th century, Cooke and Wheatstone's Telegraph, was acquired as part of an 1876 exhibition. A pioneering amateur-built x-ray set from 1896, created almost immediately after x-rays had been discovered, was acquired in 1938, and indeed was chosen as the "winner" of the competition. The other six objects were acquired after World War II; they include: two examples of wartime and early postwar technology—a V-2 Rocket and the pilot Automatic Computing Engine computer, built for the National Physical Laboratory by the computer pioneer Alan Turing. The reconstruction of fragments of the original model of the DNA double helix and the command module of the Apollo 10 spaceship were both acquired in the late 1970's. A sample of Fleming's earliest Penicillin mold and a Model T Ford were acquired in the 1990's.

It is a striking characteristic of the Science Museum that, despite changes in methods and in the particularities of display, the original message intended by such early promoters as Huxley, Lockyer, Playfair and Lyons still obtains, still rings out strongly. On December 16, 2009, the Manchester-based particle physicist Professor Brian Cox launched a scorching critique of government cuts in the funding of science; he did so via a television film made at the Science Museum. As the camera scanned the Making the Modern World gallery, his voice-over began: "This is the Science Museum. You only have to look around for a minute to understand that our wealth as a country and as a civilization was built on science and on engineering. Given the obvious importance of science in the U.K., we invest, I think, surprisingly little..." This was a view that Norman Lockyer would have been very happy to see promoted by the Science Museum he had worked so hard to create a century before.

MILESTONES OF
SCIENCE AND TECHNOLOGY

The following pages show 112 key developments
in science, technology and medicine. The purpose of the essays
and accompanying photographs is to demonstrate to as wide an audience
as possible the creative achievements that have been made from the early
Middle Ages to the present day—scientific and technological
achievements that have shaped the modern world.

BYZANTINE SUNDIAL-CALENDAR (c. 520)

The Byzantine sundial-calendar combined the typical elements of a sundial—a device that uses the sun's light to indicate time—with a geared calendrical apparatus used for showing the current phase of the moon, day of the month and places of the sun and moon in the zodiac. This sundial-calendar owes much to other inventions from both Greek and Arab antiquity, possibly even suggesting that ideas and/or technology from these innovations were somehow shared.

A common feature found in much mechanized engineering is the use of toothed gearwheels, and the early Byzantine sundial-calendar pictured here contains what are thought to be the second-oldest gears in the world. Incomplete as the device now is, and simple as it probably was when it was created, it is of immense historical importance.

The world's oldest geared mechanism, known as the Antikythera Mechanism (named for the island near which it was recovered from the sea), is in the National Archaeological Museum in Athens, Greece. Dating from the 1st century B.C., it contains an astonishingly complex assembly of more than thirty gears, the purpose of which cannot be ascertained entirely—although the nature of the mechanism indicates connections with astronomy and with the calendar.

It had long been supposed by researchers that the Antikythera Mechanism was the sole representative of a Hellenistic tradition of geared devices that, sometime in the ensuing thousand years, was transmitted to the Islamic world. The next known evidence for a geared mechanism after the Antikythera Mechanism was an account by the Persian polymath al-Biruni, in about 1000 A.D., of the "Box of the Moon," a device in which gears connected four displays showing days, "age" of the moon (days since the last new moon), and the positions of the sun and moon in the zodiac.

The discovery of the Byzantine sundial-calendar gives us a sort of reference point in the middle of the thousand-year interval between the Antikythera Mechanism and the Box of the Moon. As with the Antikythera Mechanism, its Greek inscriptions place the Byzantine sundial-calendar at least partly in the world of Hellenistic culture. And, the arrangement of the surviving wheels corresponds exactly to those described by al-Biruni, suggesting a strong probability that the technology for the Box of the Moon was borrowed, at least in part, from the Greeks. Thus, the Byzantine sundial-calendar provides the first direct evidence for the transmission of mechanical technology from Hellenistic to Arabic culture.

From the fragments, it is apparent that this Byzantine sundial-calendar has two distinct functions: that of a sundial and of a calendar. A large disk and an arm with a swiveling ring connected to the disk are recognizable as the elements of a portable sundial. In practice, this sundial was hung upright,

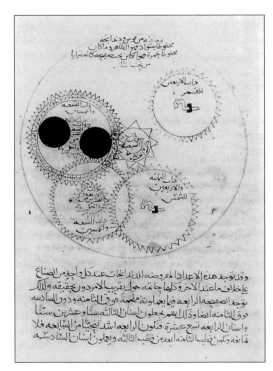

The arrangement of the wheels in al-Biruni's "Box of the Moon," from one of the earliest known manuscripts of his work, dated 1217 A.D. The central wheels and the pair with the black discs correspond to surviving parts of the Byzantine sundial-calendar (University of Leiden Library)

which explains why the arm and the disk were attached. The shape of the arm suggests that, unlike other known devices, the frame was a circular enclosure—and not just the disk by itself. That enclosure contained a geared calendrical mechanism very much like the one described by al-Biruni—which makes this instrument unique. Heads engraved in a circle on the large disk are characterizations of the days of the week, and the piece that fitted through the hole in the center of the disk was one that was most likely turned by hand, perhaps by an operator putting a key on a projecting square. Another surviving gear is connected to the days-of-the-week gear, which provides a visual display of the phases of the moon. Similar mechanisms gradually became common parts of wrist-watches, but this is the oldest known example.

This is, speculatively, about as far as the fragments can take us, except that each of the rotating pieces carries another gear wheel that has yet to be explained. The London Science Museum's reconstruction of the apparatus was based both on accounting for and making sense of the surviving features—and by adding as little as possible to them. This decision involved following almost exactly the arrangement of al-Biruni's Box of the Moon, with displays of the positions of the sun and the moon in the zodiac.

Intriguing questions remain as to whether the original instrument may have been more complicated. For example, by a fairly simple addition to the gear system, the calendar could have been made into a predictor of lunar and solar eclipses. By the further addition of scales or tables on the (missing) back of the case (in order to assist in the complicated conversions involved), the sundial and its calendrical parts could have been used together to tell time at night, by use of the moon. Fortunately, the importance of this invention does not depend on the details of any modern reconstruction. The existence of these rare fragments, and the ability to deduce the era from which they came, are enough to place them among the treasures of science.

—MICHAEL WRIGHT

ISLAMIC GLASS ALEMBIC (c. 1100)

An alembic is the essential piece of distillation apparatus used by early chemists. The liquid to be distilled is heated in a lower vessel, and the resulting vapor condenses on the inside of the dome of the upper vessel, the alembic. The condensed vapor, in a refined liquid form once again, runs down into an internal gutter and out through a spout into a collecting vessel.

Distillation was one of the arts developed by the Alexandrian alchemists (100–900 A.D.), and period illustrations, although rather crude to our eyes, show the basic shape as well as the internal gutter and the long spout. At this early date, the process was carried out on a small scale for the purification of liquids and the preparation of medicinal essences.

The traditional skills were adopted and developed by Arab chemists, but the apparatus remained essentially the same, as indeed it does today. The Persians were particularly noted for their use of distillation to produce essential oils and perfume. The Islamic glass alembic pictured here, probably dating from between the 10th and 12th centuries, is thought to be one of the oldest to survive from the West or Near East.

Distillation is one of the oldest chemical processes known to man. By the Middle Ages it was

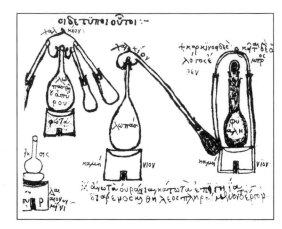

A 3rd century A.D. *illustration of alchemical distillation apparatus showing (left) two alembics*

used for the preparation of strong acids and also—perhaps a more familiar use—for making alcoholic drinks. In the late 19th century, the process was developed on an industrial scale for the distillation of petroleum, and it is still an important technique in the chemical industry.

Although analysis of this alembic confirms its authenticity, unfortunately, it is not known exactly where any of the pieces originated. It is the distinctive shape of the alembic that affords us no doubt as to its identity. Furthermore, the surface weathering is typical of Islamic glassware of the 10th to 12th centuries. The spectacular iridescence on the surface of the glass is caused by interference patterns between layers of weathered glass and the air trapped between them.

Glass-making flourished in the Islamic period, and, because of its chemical composition, the glass has a good record of survival. The basic raw materials for glass-making are silica, lime, and an alkali. In the coastal regions of the Mediterranean, marine plants that yield the sodium-based compound, soda, were used as the source of alkali. This constituent is largely unaffected by water, and it creates a glass that remains stable over time.

By contrast, the glass made in the forest glass houses, which flourished from medieval times in inland regions, includes alkali made from beech or other hardwoods. It includes potash-containing potassium, which is easily attacked by water. When buried, these glass artifacts disintegrate quickly. We are indeed fortunate to have an artifact such as the Islamic glass alembic—which has endured reasonably well over the centuries.

—Sue Cackett

THE GIUSTINIANI MEDICINE CHEST (c. 1566)

Imagine a three-month voyage on a galley in the eastern Mediterranean. Imagine falling ill with fever, dysentery, or delirium, then being cured by aromatic water of viper's bugloss, powder of chamomile, preparation of stag's horn, or metallic sulphur. The world from which the magnificent Giustiniani medicine chest survives is so remote from our own that it is difficult to empathize with the owners of what was, in effect, a huge, seagoing first-aid kit.

The chest is of solid wood construction (some of it is gilded), and its leather covering is secured by gilded nails. The chest's measurements are approximately 14 inches x 17 inches x 27 inches; it takes two people to lift it. The hinged lid is secured by a single warded lock, and opens to reveal, on the underside, a much-retouched painting on canvas of a female figure in a classical landscape. Beneath the lid, an upper compartment and three velvet-lined lower drawers, which open to right and left, contain 126 drug bottles and pots.

These drugs form a microcosm of Renaissance pharmacy, a time-capsule of medicines from the sixth decade of the 16th century, for the chest can be dated with a high level of certainty. The chest was apparently made as a medicine chest for Vincenzo Giustiniani (died 1570), the last Genoese ruler of the Aegean island of Chios. Vincenzo ruled Chios from 1562 until it fell to the Turks in 1566.

Subsequently, he became Charles II of France's ambassador to Istanbul. The arms of the Giustiniani of Chios are crudely painted on a wooden box housing scales and brass weights in the middle drawer of the chest, indicating a date of 1566 or earlier.

The 16th century saw two major changes in Western pharmacy. From antiquity, simple plant extracts had been used to make up medicines based on instructions given in the famous herbal of Dioscorides, written in Greek during the 1st century A.D. New translations of the herbal had been published in Paris and Venice in 1516, and they reached a far greater audience than had earlier manuscript versions. At the height of Dioscorides' influence, however, came new drugs from a new world—America. And later in the 16th century the remedies of the German physician, Paracelsus (1493–1541) began to be adopted. Paracelsus abhorred poly-pharmaceutical preparations (drugs mixed from a number of medicines); he introduced chemical drugs, such as sulphur and mercury, for the first time. Ninety-five of the chest's containers have paper labels with legible or partly legible names, apparently written in a 16th century hand. Of these, thirty-nine have been identified by modern scholars, and almost two-thirds of these are mentioned in Dioscorides' herbal. At least two others, guaiacum and mechiocam, are American drugs,

and another, a preparation of antimony known as *crocus metallorum*, was first described by Paracelsus. Guaiacum wood had been known in Europe since 1517: it was used as a remedy for the syphilis that raged there during the 16th century. The chest also contains extremely costly remedies, such as *terra sigillata* (sacred sealed earth from, in particular, the island of Lemnos, or sometimes Malta or Jerusalem), and bezoar stone, and at least one preparation (for preventing miscarriage) specifically for women.

The Giustiniani chest is certainly unique. It has been described as the earliest extant sea medicine chest, but it differs considerably from other sea chests. From the 16th to the 19th centuries, the chests were workmanlike kits, with instruments and bulk supplies of everyday ointments and medicines used by ships' surgeons to treat injury and disease among the crew. The nearest contemporary sea medicine chest that is known—that of the Tudor warship, *Mary Rose*, which sank in Portsmouth Harbour in 1545—would appear from its remains to have been far less sophisticated in content than the Giustiniani chest. The Giustiniani chest is perhaps best regarded as the personal medicine chest of a wealthy 16th-century family whose members would not be without it, even at sea.

—GHISLAINE LAWRENCE

THE STANDARDS OF THE REALM (1495–1601)

Shared standards of weight and measure lie at the heart of every advanced society; they promote fair and peaceful trade, the planning and construction of buildings, shipping and many other commercial activities. The search for such standards goes back to ancient times. Early peoples related measure to parts of the body—for example, the cubit (measured from the elbow to the fingertip) and the foot. The Romans adopted various weights and measures from the lands that they conquered, but certain standards that spread throughout their great empire were used in England during the period of Roman occupation, and they left their mark on subsequent British practices.

In England, there were other influences, such as those of the Anglo-Saxons. As trade grew, more measures were adopted, and local variations made the system complex and open to abuse. Successive medieval monarchs created laws to promote fair and standard measures, but enforcement of these laws was a substantial problem. Moreover, where standard measures were supplied for reference, great care had to be taken to ensure that they were made accurately and thereafter well maintained. Error could all too easily creep in, if standard measures were broken or succumbed to wear and tear.

With the Tudors came a new wave of energy. Under their rule, there was rapid economic growth in agriculture, commerce and industry. Tudor monarchs understood the need for good commercial administration, and began a vigorous campaign to establish standard weights and measures. Under Henry VII, new standards of weight, length and capacity were devised, and copies of these standards were sent to forty-three shire towns in England. Among the earliest English measures still in existence are two of Henry VII's capacity measures: the standard gallon and the standard bushel, which date from 1495.

More significant advances were made under Elizabeth I, although it proved initially difficult to establish accurate standards. A first series of weights was issued in 1558, but they proved to be too heavy. In 1574, responding to complaints that the weights in use varied too greatly, the Queen set up a committee of merchants and goldsmiths to produce new standards. Those they produced were also faulty. A second committee produced a new set of primary reference standard weights in 1582. During the next six years, fifty-seven sets of careful copies of these standards were made and sent to the major towns and cities of the realm. There they were intended to serve as a reference to be used by all local traders. Sets were also sent to the Tower of London and to the Goldsmiths' Company in London. Weights that did not match the new standards were to be destroyed, and a proclamation was issued calling for nationwide use of the new standards.

The Standards of the Realm pictured opposite are from the reign of Elizabeth I. They are made of bronze and are engraved with the royal insignia and the years of either 1582 or 1588. The troy weights (used to weigh coin or bullion) are a series of nested cups weighing from one-eighth of an ounce to 256 ounces. The avoirdupois standards (which are based on one pound equaling sixteen ounces) were used for all other general weighing, and took the form either of bell-shaped weights, with looped handles, or of a series of circular disks with slightly raised rims (to allow the weights to be stored one atop the other). They range in weight from two dram (one-sixteenth of an ounce, avoirdupois) to eight pounds.

Although standards produced by previous monarchs gradually became inaccurate and had to be withdrawn from general use, those of Elizabeth I were so good that they remained in use until 1824.

Standards of length also were issued during the Tudor era: Henry VII issued a standard yard in 1497, an octagonal measure made of bronze. The standard yard issued by Elizabeth I in 1588 was based on this predecessor. Also issued in 1588 was a standard ell, used for cloth, with a length of forty-five inches. Both of these bronze measures remained in use until 1824, from which date the ell ceased to be a legal measure.

Having issued standard weights and measures of length, Elizabeth I, in 1601, issued standard capacity measures to sixty cities in England and Wales. The bushel and gallon were to the same standard as those issued by Henry VII. The Science Museum holds a group of capacity measures of Elizabeth I: the standard bushel, two standard gallon measures, a quart, and a pint. All are made of cast bronze, with handles to use when lifting. Again, they are so well made that they were still in use in 1824.

The Tudor standards improved greatly on what had gone before and gave long service through a period in which Great Britain's might as a trading nation grew. There were still difficulties in consistently enforcing the weights and measures system: too many varieties of weights and measures remained in use (some in particular trades, some in particular localities), and legislation governing this important area was too varied to be clear and comprehensive. Yet, the Tudors had made unprecedented efforts to establish and maintain a nationwide system beneficial to everyone, at a time when a majority of the countries of Europe tended to use purely local standards of measure.

—Pippa Richardson

Until the early 1970's, the slide rule was a familiar and established tool for calculation. Although its use was taught in many schools, it was most particularly identified with the professional world of engineering. Indeed, earlier in the 20th century, the "slipstick" slide rule was virtually a symbol of the engineer, who would always have it on hand to calculate results quickly when the highest level of precision was not required.

The widespread educational and engineering use of slide rules in the 20th century disguises a much longer and richer history, for slide rules were first devised in the 17th century. The slide rule by Robert Bissaker is the earliest dated straight slide rule and is the most familiar of the several forms in which these devices have been made. The crucial ingredient is a set of logarithmic scales, so any account of slide rules must begin earlier in the 17th century with the invention of logarithms by the Scottish mathematician John Napier (1550–1617).

This new method of calculating was first described in print in Napier's *Mirifici Logarithmorum Canonis Descriptio* (*A Description of the Marvelous Canon of Logarithms*, 1614). The logarithms were taken up most quickly and actively by the "mathematical practitioners" of contemporary London. An English translation of Napier's book soon appeared, and logarithms were also mathematically reformulated in a new and more practical version of the text. The integration of logarithms into the existing repertoire of calculating devices and practical mathematics was accomplished by the English mathematician Edmund Gunter (1581–1626). Gunter published his account of a calculating instrument using logarithms in 1623, although the arrangement of his logarithmic scales was not a slide rule as such. Rather than proposing pairs of scales that would slide against one another, Gunter described a set of single scales. To use them, a pair of dividers or compasses was required, with which the quantities were stepped out along the length of the scale.

Gunter's logarithmic scales were developed in a multitude of forms—straight, circular, and spiral—by a host of mathematical practitioners and

The announcement of logarithms; frontispiece of John Napier's Mirifici Logarithmorum Canonis Descriptio, *1614*

instrument makers. The first instrument that can be called a slide rule was made public not as a straight rule but as a circular device, and considerable controversy surrounded its invention. The English mathematician William Oughtred (1575–1660), who has the best claim to have invented the circular slide rule, also appears to have been responsible for the straight rule: he created this invention by the simple expedient of placing two Gunter scales side by side.

Bissaker's straight slide rule was made before slide rules became standardized. As well as the logarithmic scales found on later navigational rules, there are additional navigational scales on Bissaker's rule which suggest that it could also have been used as a cross-staff. Compared to other surviving instruments made of wood (as opposed to the more expensive brass or even silver), Bissaker's rule is also unusual in that it has an informative inscription by the maker:

*MADE BY ROBERT BISSAKER * 1654 * FOR T W

Even with this information, however, Robert Bissaker for a long time remained no more than a name. We now know that Bissaker was an instrument maker who belonged to the Stationers' Company of London. He was of Shropshire yeoman stock, and he came to London in 1634 as an apprentice to one Nathaniel Gosse, who must have taught him the skills of the instrument maker's trade. In 1642, Bissaker was to be found making wooden instruments "at Rat-cliff, over against the Red-Lion Tavern," and there he seems to have remained into the 1660's. Presumably, he catered largely to a market of seamen and navigators: Radcliffe was a small but important town on the Thames, rapidly becoming engulfed by the eastward expansion of London.

From their very early years, particular applications were found for slide rules. New designs adapted for navigation, surveying, excise calculations, and carpentry were supplemented in the 19th century by a plethora of specialized uses, from cattle-weighing to chemistry. Despite the availability of numerical tables and mechanical calculators in the 19th and 20th centuries, slide rules remained invaluable, and by the middle years of the 20th century, they could be found in almost any profession requiring calculation. It was not until the advent of small electronic calculators in the 1970's that slide rules were comprehensively challenged and then gradually displaced. It was for good reason that the first handheld scientific calculators were known as "electronic slide rules."

Although they were never trumpeted as a grand or radical instrument destined to change the world, the portability, speed and convenience of the slide rule granted a cheap but powerful means of calculation to its owner. As devised by Robert Bissaker, the slide rule was a humble object, yet, although it lacks the glamour of more obviously heroic objects of discovery and invention, the slide rule has been essential to generations of engineers, scientists, and numerical professionals.

—STEPHEN JOHNSTON

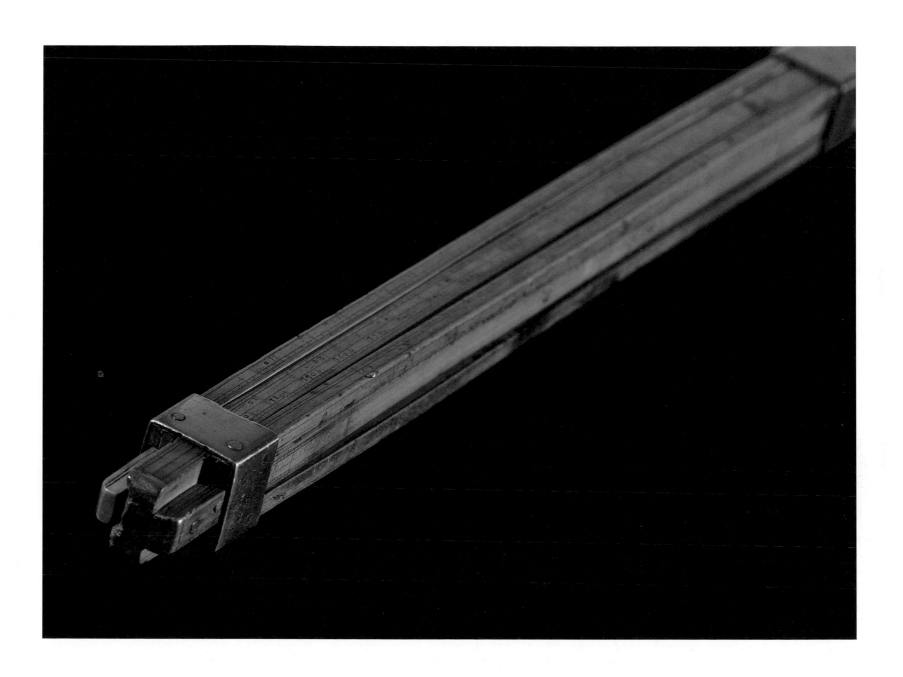

NAPIER'S BONES (c. 1690)

John Napier, Baron of Merchiston (1550–1617), is renowned for his invention of logarithms—a contribution to science viewed by some as ranking in importance second only to Isaac Newton's incomparably influential *Philosophiae Naturalis Principia Mathematica*. One of the practical advantages of logarithms is that their use allows lengthy multiplication and division of numbers to be performed using simple addition and subtraction only. The process of looking up values in printed mathematical tables and doing simple arithmetic not only simplified arithmetical tasks but also was less liable to human error than were conventional long-hand methods.

Napier's logarithms also provided the basis for the first slide rule (see the previous entry: **Slide Rule by Robert Bissaker**). In about 1620, Edmund Gunter plotted logarithms on a straight scale and performed multiplication and division by adding or subtracting lengths using a divider. The leap from Gunter's "line of numbers" to the slide rule was made by William Oughtred as early as 1621.

Napier was responsible for another intriguing physical aid to calculation known variously as Napier's Bones, Napier's Rods, or Speaking Rods, though the first of these has stuck, probably because of its macabre overtones. The "bones" do *not* refer to relics of Napier's skeleton but to the fact that the more expensive versions of the device were made from bone, horn or ivory. Napier published his description of the use of the bones in 1617, the year of his death.

The bones, devised to aid multiplication and division, consist of numbered rods or oblong blocks. Each face of the rod is marked at the top with one of the ten counting digits; below are listed each of its multiples. The rods are laid alongside each other so that the multi-digit number to be multiplied appears in the topmost row. Any multiple of this number can then be read off, right to left, along the row of the required multiple. The trick of the bones is in the layout of the numbers. The units and tens are separated by a diagonal, and in a given row the tens digit from a column on the right shares

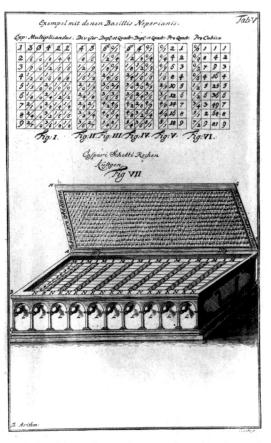

a parallelogram with the units digit in the column immediately left. By reading from right to left, and mentally adding pairs of numbers in each parallelogram on the same row, answers can be read off directly and written down. For example: to multiply 272,968 by 4, run down the left-hand index to "4" and read that row from right to left, adding number pairs—in this example, to 1,091,872.

A set of bones is in effect a multiplication table sliced up into movable columns, and the arrangement reduces multiplication to a series of look-up operations and simple additions. Division was assisted by using the bones to perform trial multiplications, and separate sets of bones were devised to calculate square and cube roots.

Variations and extensions of Napier's Bones took several forms. Sets were devised for special purposes, including geometry, planetary movements, and astronomy. Different physical arrangements were devised: there were versions with rotatable cylinders instead of oblong strips, and there was at least one version with circular disks. A final development took the form of a series of rulers devised by Henri Grenaille and which were demonstrated in 1891. Grenaille's layout removed the need for any mental effort in performing the "carry" when reading off products.

Napier's Bones were in great vogue for a time, and their popularity is evidence of the low standards of numeracy in the early 17th century, even among the well educated. Versions in varying degrees of plush were produced—numbers carved in ivory in a leather case, for example. The bones eventually lost favor as familiarity with techniques of arithmetic began to spread.

—Doron Swade

Cylindrical form of Napier's Bones, devised by Gaspard Schott, 1668. The cylinders are rotated by turning the knobs on the front of the box. Results are read off from the cylinders through slits, each of which reveals one column of figures

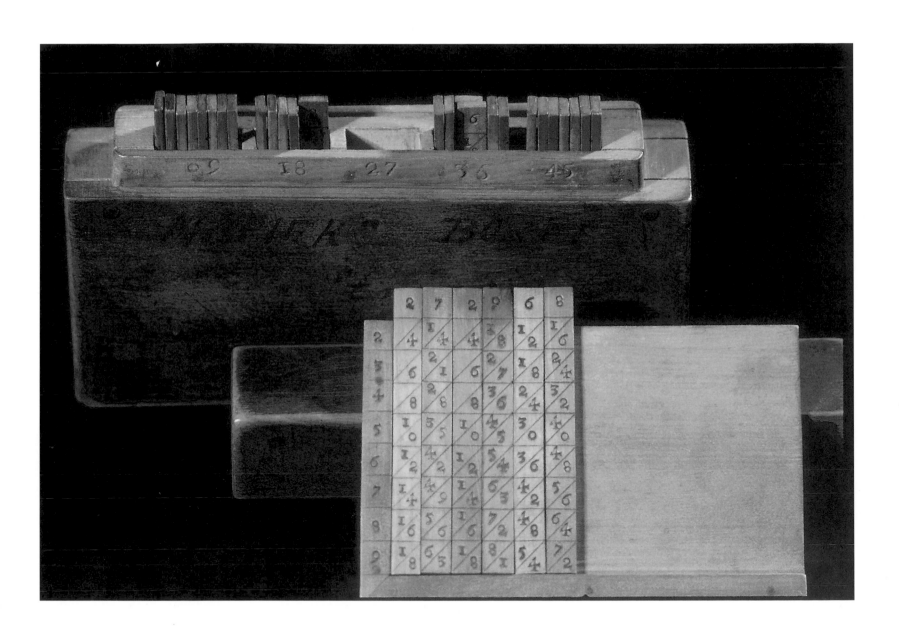

HAUKSBEE'S AIR PUMP (c. 1709)

An air pump—today we would call it a vacuum pump—is a device for extracting air from a vessel to produce a partial vacuum. In the 18th century, its importance lay in the fact that a vacuum was unexplored territory, a medium in which new experiments could be tried and unusual phenomena observed. Early pumps had been developed by Robert Boyle and others, but Francis Hauksbee (c. 1666–1713) so improved their design that his pumps were said to be the best in the world.

Francis Hauksbee first demonstrated his "New Invented Air Pump" at a meeting of the Royal Society in London in 1703, presided over by Isaac Newton, who had been elected President a few weeks earlier. The Royal Society was a forum in which men could meet to share an interest in "natural philosophy." An important rule of the Royal Society was that new ideas were demonstrated there, rather than merely discussed. "There is no other way of *Improving* NATURAL PHILOSOPHY," wrote Hauksbee in 1709, "but by Demonstrations and Conclusions, founded upon Experiments judiciously and accurately made."

Hauksbee's first demonstration of the air pump so captivated the fellows of the Royal Society that they soon appointed him official demonstrator for all of their meetings. As a professional instrument maker, he was highly proficient in the design and construction of experimental apparatus. He was also one of the first people to bring natural philosophy to a wider audience by offering public lecture courses, open to anyone willing to pay the fee. Over the years, his skillfully staged experiments and demonstrations were to provide much fodder for Newton and his contemporaries as well as the general public.

The air pump must have been made by Hauksbee some time before 1709, for he used it to illustrate his book *Physico-Mechanical Experiments on Various Subjects*, published the same year. The principle is similar to that of a bicycle pump, adapted to suck rather than blow (it takes the form of two such pumps side by side). The experimental apparatus is placed inside the vacuum vessel (*o*), which is con-

Engraving of the pump in Hauksbee's
Physico-Mechanical Experiments, *1709*

nected by a narrow tube (*h*), through valves, to the base of two cylindrical barrels. When the handle (*b*) is turned clockwise, a rack-and-pinion mechanism raises the piston (*c*) in the left-hand barrel, drawing air from the vessel down into the barrel. At the same time, the right-hand piston moves down, expelling the air inside it that had been drawn out of the vacuum vessel on the previous stroke. After a few turns clockwise, the direction of the handle must be reversed and the process repeated, with the two barrels exchanging roles. Air from the vacuum vessel is now drawn into the right-hand barrel, while air already in the left-hand one is expelled. The whole cycle must be repeated until the degree of vacuum in the vessel, indicated by a mercury barometer (*m* and *l*), reaches the required level—or until a stage is reached at which the inevitable small leakage of air back into the system balances the rate at which air can be pumped out.

A wide variety of apparatus was available for use with the pump. Some of the simpler demonstrations of Hauksbee's time, such as the one to show that the sound of a bell reduces as the degree of vacuum increases, remain in the schoolroom repertoire to this day. Others, involving living creatures, are thankfully no longer employed. Hauksbee himself made a particular study of the way that electrified objects behave in a vacuum, finding that a spinning glass globe, with the air pumped out from inside it, gives out a purple glow when rubbed with the hand. Similar experiments, done more than a century later when new sources of electricity and improved vacuum pumps were available, would eventually lead to an avalanche of discoveries that included x-rays, the electron, and the television picture tube.

Hauksbee's air pump carries special status. It links us directly with a period in history when it was a new idea that nature's secrets could best be unraveled by purposeful experiment. Neither Hauksbee nor any of his contemporaries could have foreseen just how extraordinarily fruitful that idea was to prove in the centuries to follow.

—ANTHONY WILSON

Devices that enlarge the microscopic or reduce the cosmic to a manageable human scale enjoy a wide appeal. For this reason, many other than just scientists have wished to make planetary machines that showed, using spheres of varying sizes, the relative motions and positions of bodies in the solar system.

Since the Middle Ages, planetary machines have been constructed, mostly by clockmakers, to show the relative motions and positions of the sun, moon and (sometimes) planets. Some of these planetary machines revealed astonishing complexity and sophistication in their construction, among them Giovanni de Dondi's astrarium of 1364, Philipp Imsser's astronomical clock of 1555, Eberhard Baldewein's planetary clock of 1561, and Christiaan Huygens' planetarium of 1682.

The orrery entered the scene as an altogether simpler planetary machine, and hence it was made in greater numbers and was more widely distributed than its predecessors. It is a partial model of the solar system as seen from the outside—a "God's-eye" view—with the sun at the center, according to Copernican principles. In its earliest and simplest form, the orrery showed no planets other than the earth revolving around the sun, and the moon revolving about the earth. It is in this form that the English antiquarian William Stukeley (1687–1765) described devices possibly made by Richard Cumberland and Stephen Hales. But the first serious examples were made by the celebrated London clockmaker George Graham (c. 1673–1751).

Graham probably made his planetary machines (two of which are known to survive) between 1704 and 1709, when he worked as a senior journeyman under another great clockmaker, Thomas Tompion, at the "Dial and Three Crowns" on Fleet Street. Another Fleet Street clockmaker, John Rowley (c. 1668–1728), saw Graham's planetary machines and copied one of them on a commission from Charles Boyle, fourth Earl of Orrery (1676–1731), among whose ancestors was the famous natural philosopher Robert Boyle. Although by rights the new planetary machine could well have been called a "Graham" after its inventor, or a "Rowley" after one of its earliest makers, it was in fact named an "orrery" after its purchaser and patron.

This planetary machine was the first to be called an orrery, and Rowley is thought to have made his example for the Earl of Orrery around 1712. Whereas earlier astronomical mechanical devices were generally unique objects destined for wealthy patrons, orreries were to become relatively cheap and popular during the course of the 18th century. William Jones' small portable orrery, for example, "recommends itself to the Public for Simplicity and

Internal mechanism of John Rowley's 1712 orrery

Cheapness, particularly to Masters and Governesses of Boarding Schools, Private Tutors, etc."

The original orrery is shaped like a drum with a diameter a little smaller than that of a modern bass drum. Under the central glass dome, the brass ball of the sun rotates as it ought in twenty-seven and a half earth days. The top surface of the drum, decorated with random white stars on a blue background, bears the earth-moon system under a second glass dome. One turn of the hand crank on the rim of the drum rotates the ivory sphere on the earth exactly once—a full twenty-four-hour day. The moon revolves around the earth in a lunar month of twenty-nine and a half earth days, while rotating on its own axis so as to keep the same face directed toward the earth.

Mounted on wooden pillars around the periphery is a flat brass ring representing the plane of the ecliptic—the earth's path around the sun or, equivalently, the sun's apparent path through the constellations of the zodiac. The scale on the ecliptic ring, against which a pointer attached to the rotating star-plate indicates the date, includes the symbols of the zodiacal constellations. Inside the drum are the gear-trains that control the solar, terrestrial and lunar motions.

The accuracy of Rowley's gear-cutting is simply reflected in the orbital motion and axial rotation of the three celestial bodies shown. What an orrery cannot be expected to model correctly is relative size. For the earth to be in the right relation to Rowley's three-inch sun, it would have to be the size of a grain of sand a bus-length away.

—JOHN DURANT

SISSON'S RULE (c. 1742)

At the sale of the instruments of the late ingenious optician Mr. JAMES SHORT, I purchased a finely divided brass scale, of the length of 42 inches, with a VERNIER'S division of 1000 at one end, and one of 50 at the other, whereby the 100th part of an inch is very perceptible. It was originally the property of the late Mr. GRAHAM, the celebrated watch-maker; has the name of JONATHAN SISSON engraved upon it; but is known to have been divided by the late Mr. BIRD, who then worked with SISSON.

So wrote Major General Roy in 1785. At the time, the Royal Society had in its possession another brass standard scale forty-two inches long, marked with the length of the standard yard of the Realm from the Tower, the standard yard from the Exchequer, and the French length of a half-toise, also made by Jonathan Sisson (he flourished 1736–88). Major General William Roy (1726–90) was eager to confirm that his scale matched that of the Royal Society because he intended to use his scale to prepare some of the measuring equipment for setting out a baseline on Hounslow Heath, near London. This work was a necessary preliminary to fixing the relative positions of Greenwich and Paris observatories with greater accuracy than ever before.

The principle was simple: if the length of the baseline was known, the distance between each end of the baseline and a distant landmark could be worked out using a theodolite (a surveying instrument used to measure vertical and horizontal angles; see **Ramsden's Three-Foot Theodolite**) in order to measure the angles of an imaginary triangle formed by the baseline and the landmark. A simple trigonometrical calculation gave the distances. From Hounslow Heath, a web of triangles was surveyed across Southern England toward Dover and a second baseline at Romney Marsh, and then

Telescope used at Hounslow Heath in 1784 by Major General William Roy of the Royal Engineers

simultaneous sightings across the English Channel linked the English measurements to those of the French. The work was carried out at the behest of the Royal Society, of which Roy was a Fellow, and the French Académie des Sciences. Although Roy died in 1790, his work became a model for the extension of the project to map the rest of Britain.

On April 16, 1784, a select group of scientists from the Royal Society convened at Hounslow Heath for a preliminary survey. The heath was ideal: it had an extremely level surface, was large, near both London and Greenwich, and with few obstructions likely to make measuring difficult. The team readily settled upon a line running for about five miles from the northwest extremity of the heath toward the southeast, and they agreed that soldiers should be employed to clear furze bushes and ant-hills along the track of the base and to assist with the measuring operations.

It took until early July to clear the route for the line, during which time Jesse Ramsden (1735–1800), the instrument maker, prepared the instruments. Sisson's rule (or "ruler") was a vital piece of equipment. Carefully transferring measurements

from it, using special beam compasses, the team marked out a twenty-foot master scale. This scale was next used to set a twenty-foot beam compass, which was used to mark out the instruments. These included a steel chain one hundred feet in length and three twenty-foot-long wooden rods with a standard metal rod for comparison. Measurements would be made with both, to ensure that the results would be more accurate.

The chain proved to be even more accurate than the team had hoped, and it was remarkably quick and easy to use. Two different methods of measurement were attempted with the wooden rods, one of which (putting the rods end to end) was found to be much more accurate than the other. Although wooden rods had been used to measure most of the other baselines in other countries, this method proved unsatisfactory, as humidity changes in England caused large variations in the lengths of the rods throughout the course of each day. When it became clear that this was the case, Roy took up a suggestion from Lt. Colonel Calderwood that glass tubing might be used instead, and a week of rainy weather was devoted to the construction of a suitable number of glass tubes and converting them to measuring rods. The twenty-foot master scale was rechecked using Sisson's rule and then used to check the length of the glass rods.

In the end, the base was measured using both glass rods and the one-hundred-foot chain. Work started in August 1784, and on August 21st King George III came out to Hounslow Heath to see how it was progressing. By the end of August, the base-line had crossed thirteen roads and two rivers, and it was calculated finally to be 27,404.7219 feet in length.

—JANE INSLEY

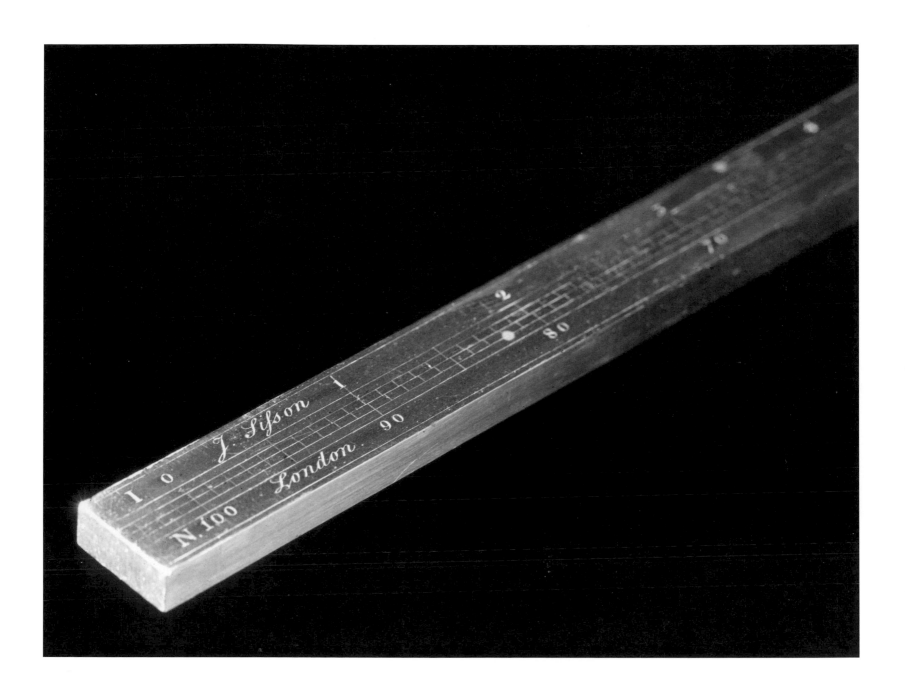

SHELTON'S ASTRONOMICAL REGULATOR (c. 1768)

In the 1760's, John Shelton (he flourished 1712–77), clockmaker in Shoe Lane, London, supplied five astronomical regulators to the Royal Society and to the Board of Longitude. These regulators were specifically intended to time the transits of Venus that took place on June 6, 1761 and on June 3, 1769, but they had a long and remarkable subsequent career. All five still exist.

An astronomical regulator is an exact clock used for timing astronomical events. The most important components are the escapement and the pendulum, and Shelton made use of two of these innovations of 18th-century horology. The "dead-beat escapement" had been pioneered by his former employer, George Graham (see **The Original Orrery**), in about 1720, and it was so accurate that it continued to be used in precision clocks for another two centuries. The "gridiron pendulum," invented by chronometer-builder John Harrison (1693–1776), counteracts the effects on the clock of changes in temperature—particularly important for these clocks, which were to travel from the Arctic to the Tropics.

Shelton's astronomical regulator pictured opposite can be described like this: the clock features a typical "regulator dial" of which the outer dial measures the seconds, the smaller dial inside measures the minutes, and the dial segment seen beneath the center measures the hours. This was a convenient arrangement for astronomers, who had to identify at a glance the exact second of an observation.

The pair of transits of Venus in the 1760's aroused immense interest all over Europe. This rare phenomenon, when the planet Venus is seen passing across the sun, could be used to measure the distance of the sun from the earth—if only it could be accurately observed at widely separate places. In one of the earliest examples of international scientific cooperation, learned institutions rose to the challenge and initiated many programs of observation.

All five of the Shelton regulators took part in one or another of the British transit expeditions, one of them being used for observations as far away as Tahiti, during Captain James Cook's first voyage (1768–71). It is not now possible to distinguish the early history of each instrument, but it appears that the regulator pictured opposite was one of those that accompanied the transit expedition in 1769 to the far north of Norway (Captain William Bayly to the North Cape, and Captain Jeremiah Dixon to Hammerfest). A year or two later, it seems to have been one of the two precision clocks taken by Captain Cook on his second and third voyages of discovery to the South Seas (1772–75 and 1776–80).

One of the five Shelton regulators was used in one of the most famous experiments of the 18th century, Nevil Maskelyne's measurement in 1774 of the weight of the earth. He used it to see how much a pendulum's rate was influenced by a neighboring mass—Mount Schiehallion in Scotland. Many years later, in 1828, the regulator pictured opposite was used by the Astronomer Royal, George Airy, in another experiment to weigh the earth—comparing the rate of a pendulum at the top and the bottom of a deep mine, Dolcoath copper mine in Cornwall.

The Shelton regulators found their fullest use in helping to solve yet another important astronomical problem: measuring the exact shape of the earth. The rate of a pendulum depends on how far it is from the center of the earth. It will swing more slowly at the top of a high mountain than it will at sea level. If enough measurements are made worldwide, the shape of the earth can be established. The Shelton regulators were used for these "pendulum" measurements as early as the 1760's, and they later accompanied many expeditions all over the world.

The regulator pictured opposite was with the British scientist Edward Sabine on four separate pendulum expeditions to the North and South Atlantic (1818–23), reaching as far north as the 79th parallel in Spitsbergen, north of Norway. In 1865–73 it travelled the length and breadth of India with Captains J.P. Basevi and W.J. Heaviside of the Great Trigonometrical Survey of India. In 1882–84, on loan to the United States Coast and Geodetic Survey, it saw service in Australasia, Singapore, Japan, and the United States. In Auckland, the Americans used it for the transit of Venus that took place in 1882—113 years after the famous transit of 1769 for which it had originally been constructed.

Not much is known about Shelton's life. He was apprenticed in 1712, and later he worked for George Graham, one of the most distinguished of 18th-century clockmakers. He was Graham's chief constructor of astronomical clocks. But in 1777, then in his eighties, he was destitute and presented a petition for assistance to the Royal Society "setting forth that a great many clocks and astronomical regulators have been made by him…which are in position in the chief observatories of the world…" The outcome of this petition is not known.

The art of precision clockmaking was well developed by the second half of the 18th century, and Shelton's regulators do not present any significant innovations of their own. But they are remarkable for their long life: there can be few scientific instruments that have travelled so far over such a long period for such a variety of uses.

—GRAEME FYFFE

ARKWRIGHT'S SPINNING MACHINE (1769)

Perhaps the most visible and dramatic memorial of industrial capitalism as it was pioneered in Britain in the second half of the 18th century is the factory. It was a new and specialized kind of building designed not only to make efficient use of the new forms of machine (to which water or steam power could be applied), but also to serve as a new setting for the management of labor. Indeed, factories became more than merely buildings; they became the new system of industry.

The factory—or mill, as it was known in the textile industry—became a potent symbol of the triumphant power of industrial production—as well as a symbol of the oppression and degradation of workers. It was the focal point of the new industrial towns and the sole source of their great wealth and prosperity—but it also became the scene of violent machine-breaking by those who saw their livelihoods threatened. Paradoxically, in recent years it has been the closure of factories that has been blamed not just for the extinction of jobs but also for the destruction of the social spirit that many of those industrial communities had cherished for generations.

The first factories (in the form that we have come to recognize them) were built to house new machines for spinning yarn. These machines were introduced into a business that had hitherto been conducted on a domestic, rather than an industrial scale. Arkwright's spinning machine, and the factory organization he introduced to exploit it, changed all that—suddenly and dramatically. That change in preparing fibers and spinning yarn soon spread to other branches of the textile industry, and in turn the textile industry set a pattern for other industrial enterprises.

Sir Richard Arkwright (1732–92) had little or no schooling; he began his career as a barber. Later he travelled, collecting hair for wig-making, and those travels may have given him the opportunity to hear of the numerous efforts then being made to improve production in the textile industry, attempts on which he was to base his own ideas. He was probably not much of an original inventor, but he was exceptionally good at seizing on a promising idea and exploiting it.

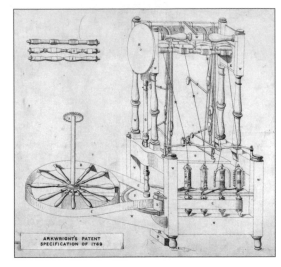

The published lithograph of a drawing accompanying Arkwright's patent specification of 1769

Arkwright's forceful approach made him many enemies, but he became a wealthy man. He was knighted in 1786, and was appointed High Sheriff of Derbyshire in 1787. His portrait, by Joseph Wright of Derby, makes him appear smug, arrogant and powerful—not an engaging man. Matthew Boulton, a contemporary industrialist of a very different stamp, wrote of him: "Tyranny and an improper use of power will not do in this country… If Arkwright had been a more civilized being and understood mankind better, he would now have enjoyed his patent."

In spinning, fibers are drawn out ("drafted") and then twisted together. Arkwright's machine used the "flyer" of the conventional hand spinning wheel to provide the twist. But whereas the hand spinner drafted the fibers by teasing them down from the distaff, the action of the fingers, in Arkwright's machine, was simulated by pairs of "drafting rollers." There were two pairs of these for each thread. The second set was driven slightly faster than the first so that the fibers were teased out between them.

This use of rollers was not new. It may be traced back to Lewis Paul's patent of 1738. But Arkwright's

machine was better arranged. The small machine pictured opposite resembles so closely the drawing of Arkwright's patent of 1769 that it is likely to be his prototype or development model from that time. It bears the scars of having been modified, as one would expect. The machine originally had four spindles, and it might have been suitable for use in the home, according to the old domestic system of textile manufacturing. In fact, that kind of use was not to happen: Arkwright moved directly to the grouping of many spindles together under one roof, to be driven by a central source of power.

His first mill, at Nottingham, was driven by a horse. That mill has long since disappeared. But, in 1771, Arkwright and his partner built a larger mill, powered by water. This mill was at Cromford in Derbyshire; it survives, and is now a museum. Arkwright sold licenses to a few other spinners, permitting them to use his machine, but it seems certain that the water frame, as it came to be called, was widely pirated in the 1780's. There are many remains of early spinning mills, especially in the north Midlands of England, and their remarkable conformity to the standard width of approximately thirty feet, set by Arkwright's own factories, suggests that they were built to house water frames. After 1790, when mule spinning was gradually adopted, new cotton mills were built fifteen or more feet wider to accommodate the new machines.

Arkwright licensed his innovation only in units of a thousand spindles, thus forcing licensees to adopt a system similar to his own. He seems to have done this primarily in an attempt to control the use of his patents; but the effect of grouping together large numbers of machines, sharing a central power source, and having an organization and a division of labor to attend to them, was significant in determining the pattern of development of the factory system into the 19th century.

—Neil Cossons

TROUGHTON'S DIVIDING ENGINE (1778)

In the 18th century, the scales on navigational instruments were marked out by skilled craftsmen in a process called "dividing." Thus, accurate instruments took a long time to make and were quite expensive. From the 1770's onward, dividing was partially mechanized. Dividing engines were introduced, enabling instrument makers to satisfy a growing demand for improved navigational instruments. At first, they were capable only of dividing small instruments, but by the 1850's dividing engines were being used to graduate the scales on large astronomical telescopes, and hand-dividing was obsolete.

The dividing engine pictured opposite was completed by John Troughton (c. 1739–1807) in 1778, and it is important because it is similar to the first successful such machine made, completed by Jesse Ramsden (1735–1800) in about 1775.

The incentive to develop these tools came from a major navigational problem: how to find a ship's longitude. In the 18th century, Britain relied on its navy for defense and its merchant ships for trade, so accurate navigation was vital. The British Government set up the Board of Longitude in 1714, and this board offered a top prize of £20,000 (equivalent to at least £1 million today) to anyone who could devise a way to find longitude to within thirty nautical miles.

By the 1760's, John Harrison and Tobias Mayer had developed practical methods of finding longitude at sea—the former, by building an accurate chronometer; the latter, by tabulating the motion of the moon. Both men were rewarded by the Board of Longitude. Their methods needed accurate angle-measuring instruments (such as the octant and sextant) to measure the altitudes of heavenly bodies, and this need spurred a demand for these instruments, but accurately dividing such small scales by hand slowed production.

The English instrument maker Jesse Ramsden provided the solution: he mechanized dividing. Clockmakers had used machines to position and cut gear-teeth since the late 17th century. In about 1740, a York clockmaker, Henry Hindley, built a more accurate version, which could also be used

Ramsden's second dividing engine; engraving from Description of an Engine for Dividing Mathematical Instruments *by Jesse Ramsden, 1777*

to divide instrument scales. Hindley's machine was the earliest dividing engine, and it incorporated the worm-and-gearwheel arrangement used in later engines. It was used to divide a few astronomical instruments, but it remained little known. One of his workmen, John Stancliffe, who later became his foreman, may have told Ramsden of Hindley's machine. Ramsden's first dividing engine was completed in about 1766, but it was not accurate enough for navigational and astronomical instruments. His second engine, completed in about 1775, was a complete success. The Board of Longitude appreciated its importance and wanted to make it available to all instrument makers. Ramsden was paid £300 for disclosing its secrets, and the board bought the engine for £315, at the same time allowing Ramsden to retain and to use it. As

a condition of the payment, Ramsden, between October 1775 and October 1777, was to train up to ten other instrument makers in how to duplicate and use the engine. Also, he was asked to publish a description of the engine, with enough detail to allow other instrument makers to follow his instructions in order to duplicate it.

John Troughton was probably one of the instrument makers taught by Ramsden. He had an excellent reputation as a hand-divider, and he was well placed to appreciate the importance of this new tool. In 1778, he completed his dividing engine, which took him some three years to build. As Ramsden's book on the dividing engine was not published until 1777, it is difficult to imagine how else John Troughton could have obtained full details of Ramsden's work, except by being one of his students.

Troughton's engine was profitable. Ramsden had a reputation for long delivery times and occasionally bad workmanship: instrument makers preferred to pay more for a quicker dividing service from John Troughton, who was said to have made more money with his engine than Ramsden received from the Board of Longitude. It earned him enough to purchase the instrument-making shop of Benjamin Cole at 136 Fleet Street, London, in 1782, later to form part of the famous firm of Cooke, Troughton and Simms.

Troughton's dividing engine was simple to operate. The instrument being divided was fixed to a large wheel on top of the engine. When the treadle was pressed, the wheel and the instrument were turned through a fixed angle. Then, with his right hand, the operator used a cutting tool guided by a system of swinging links to mark the instrument scale. The process was repeated until the complete scale had been divided. Although this process was many times faster than hand-dividing, it was a back-breaking task to lean over the engine to work on a small sextant. John's brother Edward wrote that "it had done no good either to his health or that of my own, and had materially injured that of a worthy young man, then my assistant."

—PETER TURVEY

HERSCHEL'S SEVEN-FOOT TELESCOPE (c. 1781)

During the 18th century, few people had a greater impact on astronomy than did William Herschel (1738–1822). At a time when astronomers were preoccupied with measuring the positions of stars to ever greater levels of precision, he was busy exploring the structure of the universe. He was also active in advancing telescope technology.

Herschel's serious interest in astronomy did not develop until he moved from Hanover, Germany to England in 1773. He was then a professional musician and composer, and he had gained the post of Director of Music for Bath. In his spare time he developed his interest in astronomy, and, having found commercial telescopes unsatisfactory, he learned how to make his own. Herschel built his telescopes to a design first suggested by Isaac Newton, which used mirrors instead of lenses to focus light. In time, he perfected his instruments until they were superior to any others then available. The excellence of his telescope (similar to the one pictured opposite) led to his discovery, in 1781, of the planet Uranus. This discovery made Herschel famous overnight: no new planets had been observed since antiquity. He originally named the planet "Georgium Sidus" (George's Star), in honor of the reigning King George III, but fellow astronomers did not agree with his choice, and the name was changed to the one we know now.

In 1782 Herschel was appointed "Royal Astronomer," and this royal patronage allowed him to give up his musical duties and move to Datchet, near Windsor, where he could devote himself fully to astronomy.

Herschel made and sold many telescopes during his long career. This example—Herschel's seven-foot telescope—was made for his sister, Caroline Herschel (1750–1848): it was a present from him. Caroline was a very capable astronomer in her own right; she acted as William's assistant when he was observing. Her contributions to astronomy have been somewhat overshadowed by those of her more famous brother. She is best known for her careful observations of comets: between 1786 and 1797 she discovered a total of eight comets. This was a

Caroline Herschel, English astronomer, 1841

remarkable achievement for the time; even today, it is impressive. She also undertook the onerous task of correcting the numerous errors in Flamsteed's star catalogue, *Historia Coelestis Britannica*, the standard work of its day, which William used in his searches of the heavens.

The telescope that William used to discover Uranus was like the one pictured opposite. It had a six-inch-diameter mirror with a focal length of seven feet. (The image is formed seven feet from the mirror.) The mirror was made of speculum metal, a highly reflective but very hard alloy of copper and tin that can be ground or polished only with great difficulty. Unfortunately, speculum mirrors easily tarnish and therefore must be re-polished frequently. The design of this telescope is typical of instruments made throughout Herschel's career, and it had an influence on telescope makers for more than a hundred years. Astronomers needed larger telescopes to see fainter objects and to distinguish more detail. Herschel showed them how to make these larger telescopes at a cost that was modest compared to other designs then available (*see* **The Rosse Mirror**).

Using the experience that he had gained with his smaller telescopes, Herschel was able to construct two larger ones that had focal lengths of twenty feet. One was built while he was still living

in Bath, but, because of the small size of his garden where it was placed, it proved difficult to use. After his move, and using his new instrument, Herschel embarked on a systematic survey of the heavens, and this was to prove the most productive period of his life. He was able to confirm his earlier observations while making new discoveries. He was also able to uncover many more double stars, some of which he proved were rotating around each other. There had been a long debate during the 18th century as to why these stars appeared double. Some astronomers thought that this phenomenon was simply a "line of sight effect," that the stars lay in the same direction but were at different distances from the earth. Herschel's observations over many years proved that at least some of these stars moved in orbit around each other.

These successes spurred Herschel to build an even larger telescope, and, with patronage from the King, he was able to cast, polish and grind several speculum mirrors of forty-eight inches in diameter. The resulting forty-foot telescope was erected at Herschel's new home in Slough in 1789, where, because of its sheer size, it became an instant landmark. During its construction the King was a frequent visitor, and he delighted in conducting people through the enormous telescope tube. On one occasion, while he was with the Archbishop of Canterbury, the King was heard to say, "Come, my Lord Bishop, I will show you the way to Heaven!" However, the telescope was not as successful as Herschel had hoped, and it was little used: this failure was partly the result of its huge bulk—it took three to four people to operate. The instrument fell into disuse in 1815, and it was finally dismantled in 1839.

On William's death in 1822, Caroline Herschel returned to Hanover, where she lived until her death at the age of ninety-seven. Such was her fame in her later years that she was awarded a gold medal by the Royal Astronomical Society. She was also made an honorary member of the Royal Society, the first woman to receive this accolade.

—KEVIN JOHNSON

BOULTON AND WATT ROTATIVE ENGINE (1788)

"I sell here, Sir, what all the world desires to have—power!" was the proud boast of Matthew Boulton (1728–1809) to a visitor to the famous Soho Foundry in Birmingham, England. What Boulton was selling, with his partner James Watt (1736–1819), was the power of steam, the single most important technological factor in bringing about the huge economic and social changes that we now call the Industrial Revolution. Steam drained mines, powered machinery, propelled locomotives and ships: it had helped transform Great Britain and many other countries into highly industrialized societies. Even today, although the older reciprocating engine has been largely replaced, it is still steam power—in the form of turbine-driven alternators producing electricity—that satisfies a large proportion of the world's energy needs.

James Watt did not invent the steam engine, and his engines did not lead to the emergence of the factory system, which was pioneered initially on water power. What he did contribute, however, were outstanding improvements in efficiency and an engine that would progressively supplant the water wheel, thus freeing factory owners from the geographical and seasonal limitations of having to place their factories in steeply graded river valleys.

Thomas Newcomen (1663–1729) had built the first recorded engine in 1712, and the machine was well established in its use for mine drainage before Watt was born. When he was still a young instrument maker, Watt was asked to create an improved version of Newcomen's engine. Watt's appreciation of the inherent inefficiency of the Newcomen engine—the thermal losses resulting from raising and lowering the temperature of the working cylinder during the cycle from steam inlet to condensation—led him to a brilliant solution: the separate condenser—the greatest single improvement ever made to the steam engine. By changing the location of the condensation from the cylinder to a separate evacuated chamber (so that the temperature of the working cylinder was not significantly low-

ered), he brought about a major increase in thermal efficiency and in absolute power.

Watt patented his condenser in 1769. In 1782, he introduced the double-acting engine in which steam was applied alternately to both sides of the piston, thereby obtaining double the power from the same size cylinder. In the same year, he took out a patent for a rotative beam engine in which the beam drove a shaft and flywheel. These features, together with his parallel motion (which provided a positive linkage between the reciprocating end of the piston rod and the beam), meant that Boulton and Watt had developed an engine capable of driving machinery. By 1800, when Watt's partnership with Boulton ended, and the patent on the separate condenser had expired, 451 engines had been built, of which 268 were rotative.

The Boulton and Watt rotative engine pictured opposite was built in 1788 to power a section of Boulton's own works: it incorporates all of the features developed to that date. The connecting rod

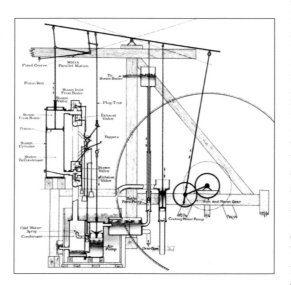

A diagrammatic sectional view of the 1788 Boulton and Watt rotative beam engine

is linked to the flywheel shaft by way of sun-and-planet gearing, possibly devised to avoid infringing James Pickard's patent for the common crank. It is also the first engine ever fitted with a centrifugal governor to regulate its speed.

The engine proved the reliability of its design and construction by giving seventy years of good service. As with any machine that had been so long in use, there were repairs and the replacement of parts, but it is believed that this is the oldest engine in the world that is both substantially complete and in its original state. At the time of its construction, Boulton and Watt still did not generally build whole engines. They followed the practice of providing a design and then advising about the sources of supply of major parts or materials. They would offer a trained erector to oversee the building of the engine, and they usually provided some critical components such as the "nozzles" or valve chests. Therefore, no two engines were quite alike.

The 1788 engine not only marks the introduction of steam power to a wide range of industrial applications, but it also reveals a number of general truths about innovation and technology transfer. Boulton and Watt dominated the market after 1780, but that was partly the result of their energetic defense of their patents. Watt became progressively more conservative: he was opposed to high-pressure steam engines, and he was reluctant to consider steam for road or rail traction or steam engines for boats. Although by 1800 Watt's engines were still the most reliable—largely as the result of superior design and workmanship—they were also old-fashioned. Scores of other engineers were pioneering new developments. By the end of the 18th century, approximately 1,200 to 1,300 steam engines had been built, nearly 500 of these based on Watt's designs. Before the end of the 19th century, the steam engine had become the universal source of power for industry and transport.

—NEIL COSSONS

SYMINGTON'S MARINE ENGINE (1788)

The mechanical propulsion of ships is a challenge that has fascinated experimenters since the Renaissance. They would all have been thoroughly familiar with the waterwheel, the origin of which is known to predate the Roman era. The waterwheel was thus an obvious prototype to be copied in trials of the mechanical propulsion of ships. Among the many 18th-century figures attracted by this quest was a wealthy English banker, Patrick Miller (1731–1815), who commissioned a series of double- and triple-hulled vessels with paddle wheels between their hulls: these wheels were turned by manual capstans. Human stamina, however, proved insufficient to propel any of these boats for more than a few minutes. Miller was about to discontinue his experiments when he learned from the tutor he had engaged to educate his sons about a twenty-year-old engineer named William Symington (1763–1831), who was then working in the Lanarkshire, Scotland lead mines. Symington had successfully built an experimental beam engine that combined the thermal efficiency of a separate condenser with the simplicity of an atmospheric engine based on Thomas Newcomen's principles. In 1786, Patrick Miller witnessed a demonstration of a steam carriage fitted with one of Symington's engines. When Symington was granted a patent the following year, Miller was sufficiently confident of the engineer's potential that he ordered an engine to power a catamaran on the lake at his Dalswinton estate in Dumfriesshire.

Symington's response was a wooden-framed atmospheric engine, the power of which came from the vacuum created when a cylinder full of steam was suddenly condensed. Because each of the two vertical pistons created power only on the down stroke, these pistons were linked by chains to a drum suspended above them, which revolved in alternate directions as the pistons rose and fell. The machine's principal ingenuity lay in Symington's use of ratchets and pawls (previously patented by Matthew Wasbrough, in England only, in 1779) to convert semi-circular motion into continuous rotation. Chains connected the working drum with a pair of loose pulleys on each of the two paddle shafts. Ratchet teeth around the interior of each pulley engaged alternately with opposite facing pawls keyed to the two paddle shafts. Interconnection between the sets of pulleys ensured that both paddle shafts rotated continuously in a uniform direction as the pistons rose and fell in their open-topped brass cylinders. This ingenious alternative to the crank previously patented by James Pickard is claimed to have propelled Miller's steamboat at five knots in trials on Dalswinton Loch. Like many other claims advanced by early steam engine promoters, this claim seems highly exaggerated.

Whatever the vessel's performance, Miller was sufficiently encouraged to commission a much larger engine of almost eleven horsepower from

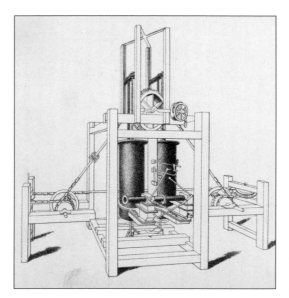

Lithograph of the engines of Patrick Miller's experimental steamboat of 1788

his protégé, and Symington's next engine successfully propelled a sixty-foot catamaran on the Forth and Clyde Canal, near Falkirk, in December 1789. But at this larger scale the ratchet arrangement revealed its inadequacy for sustained power transmission under load, and the pawls failed repeatedly. A curt rebuff from James Watt to Miller's proposal for a collaborative development prompted the Edinburgh banker to give up steam engineering and to devote his energies to agricultural improvement; Miller was also apprehensive of litigation by Boulton and Watt over alleged infringements of the separate condenser patent. He removed Symington's smaller marine engine to his library, where it remained as a curio.

This failure was not, however, the end of Symington's involvement with steamboats. In 1801, he produced a different kind of engine for the experimental canal tug *Charlotte Dundas*; however, the Forth & Clyde Canal Navigation Company decided not to use steam tugs, as their wash would damage the canal banks.

Miller's son inherited the engine on his father's death in 1815. After years of neglect, he decided to sell it for scrap. An Edinburgh plumber got as far as discarding the worthless and wooden framework and separating the non-ferrous components before he died. In 1853, agents for Bennet Woodcroft (1803–79), himself a pioneer of steam navigation, located the surviving components in the plumber's workshop and purchased them for London's Patent Museum.

Predating as it did Henry Bell's *Comet* steamship (*see* **The *Comet* Steam Engine**) by more than twenty years, the engine conceived by William Symington to advance the progress of steam power, despite antagonism from James Watt and his colleagues, certainly proved itself to be one of the forerunners in the field of marine engines.

—JOHN ROBINSON

It is occasionally the case that the peak of achievement in one area of science or engineering ironically becomes the starting point for a new, quite different one. The three-foot theodolite of 1791, by Jesse Ramsden (1735–1800), is an example of this phenomenon.

The theodolite (a surveying instrument used for measuring vertical and horizontal angles) remained in constant use for more than sixty years. It was used for measuring the angles in the Primary Triangulation of Great Britain, the foundation for the first Ordnance Survey maps of the country. This huge undertaking was initiated by Major General William Roy (*see* **Sisson's Rule**) with a trigonometrical operation to fix the relative positions of the observatories of Paris and London more accurately than ever before—indeed, the work was done so accurately that it would be 150 years before the measurements were done again. The first stage was the measurement of a length of a baseline on Hounslow Heath (in what is now Greater London) in 1784. One end, still marked by a cannon, is a national monument lying within the boundary of London's Heathrow Airport.

The survey was to reach across England toward Dover; then, sightings across the Channel would link the English and French surveys. The scheme, suggested by Cassini de Thury through the French Ambassador to King George III, was given to the Royal Society to implement, with royal funding. Sir Joseph Banks (1743–1820), the President of the Royal Society, knew that William Roy had argued persuasively for more than thirty years to set up an accurate mapping framework nationwide, and Banks asked him to undertake the work. Roy leapt at the challenge, as he felt that the project could well develop into the national triangulation he had long advocated—provided that the work was of the highest quality.

Jesse Ramsden was one of the top ten specialist instrument makers in 18th-century England: he supplied large-scale astronomical and surveying instruments to clients both at home and abroad. He became a Fellow of the Royal Society in 1786.

Ramsden's first three-foot theodolite was made for the Royal Society, and it took him three years to complete it. When it was finished in 1787, Roy could take a bearing on a mark at up to seventy miles away, with an error in the angle-reading of only 1/180th of a degree: it was capable of measuring the curvature of the Earth.

The second instrument made by Ramsden (pictured opposite), was even better, and the accuracy of its scales was exceeded only in the 1930's. It was

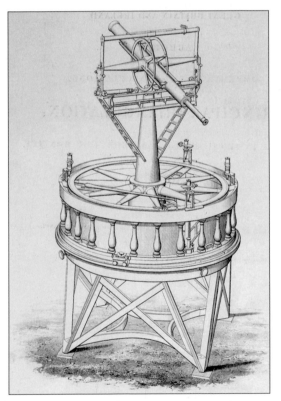

Ramsden's Great Theodolite: frontispiece to Account of the Observations and Calculations, of the Principal Triangulation, *1858. The horizontal circle, carrying the scale, was the most important part of the instrument, and it needed to be kept as flat and immobile as possible. The instrument had to withstand a great deal of wear and tear in use, transport and weather.*

commissioned by the Survey of India, but Ramsden incorporated improvements to the design that increased the cost to £373, more than the Survey of India was prepared to pay. Fortunately for Ramsden, the third Duke of Richmond, Master General of the Ordnance, agreed to purchase his white elephant, and the Primary Triangulation of Great Britain was formally authorized with the sanction of King George III.

Both theodolites were used for the most important of the 229 stations in the network, and, despite their enormous weight, they were transported to the tops of mountains, steeples, or even scaffolds specially built to accommodate them, so that the lines of sight could be maintained.

In installing the theodolite before taking a measurement, the experts immediately encountered a problem: how to provide an entirely solid foundation. Solving this problem occasionally required digging: in Holme Moss, nine feet of bog and six feet of sand had to be removed to get down to a steady base. On exposed mountains tops, a wall of stone would be built around the base of the observatory tower. This precaution certainly saved the instrument one day in 1838, when a storm moved a temporary observatory out of position, even though the protective wall was two feet thick.

The names of the observers and the "bookers" (the men who wrote down the readings in the observation books) were recorded at every station. All the calculations and subsequent map drawings were carried out at the office. For the first fifty years of the Ordnance Survey, the office was at the Tower of London. After a disastrous fire in 1841 in a building next to the Ordnance Map Office, the Survey moved to Southampton. Nearly one hundred years later, the first Ramsden theodolite fell victim to a German air attack, during which the building was almost completely destroyed. The Board of Ordnance theodolite is therefore the oldest of its kind to survive.

—JANE INSLEY

MAUDSLAY'S SCREW-CUTTING LATHE (c. 1797)

The successful construction of all machinery depends on the perfection of the tools employed… The contrivance and construction of tools must therefore ever stand at the head of the industrial arts. —Charles Babbage, 1851

Henry Maudslay (1771–1831) began work at twelve years of age, filling cartridges at Woolwich Arsenal; eventually, he became a highly skilled blacksmith. At eighteen, he was recommended to inventor and locksmith Joseph Bramah as the man to solve problems in the manufacture of Bramah's lock, and Maudslay worked for him until 1797, when he set up on his own. From the outset, Maudslay always strove to work "in the best possible manner," and his lathe, designed specifically for making screws, reflects his commitment to perfection.

Screws are vital components in machinery. They are used not only as fastenings but also as a means of adjustment. For these purposes no great precision is usually demanded, but it is convenient to have them well-fitted, smooth in action, and preferably interchangeable with one another. But screws undertake more demanding roles with their use in measurement and in providing regular, controlled motion in machinery: for these purposes, a much higher standard of precision is needed. Screws were usually made by very crude methods, so these properties could not be taken for granted.

The modern method of making a screw with any pretension to accuracy is to rotate a cylindrical workpiece in a lathe, while a cutting tool, held by a slide rest, is automatically moved along to form a helical groove in the workpiece. Usually, the slide rest is moved by a leadscrew geared to the rotating workpiece. Maudslay was not the first to use either the slide rest or the leadscrew, both of which are essential parts of his machine. However, it is certain that his remarkable achievement was to have introduced such tools into everyday use.

Henry Maudslay: lithograph by H. Grevedon, 1827. Photographed from the original portrait on stone in the collections of the Science Museum

Maudslay's screw-cutting lathe is built on two parallel triangular bars. One bar carries conventional headstocks to hold the workpiece. A slide rest riding on both bars carries a tool-holder that is equipped with a screw feed; it includes a micrometer dial to regulate the depth of the cut. The leadscrew carried between the bars moves the slide rest along; it is so mounted that it may be readily changed. The gears that connect the leadscrew to the mandrel rotating the workpiece are lost, as is the means of driving the machine, a capstan wheel worked by hand.

In 1846, the famous lathe manufacturer and machinist Charles Holtzapffel (1806–47) offered a detailed description of how the lathe was used. The problem that faced Maudslay was how to originate an accurate screw thread without copying the errors in a pre-existing master. His technique was to originate a new screw thread by using a tool with an inclined knife, which cut a helical trace on a soft metal rod. Selected screws were used as leadscrews in the lathe to cut new screws in any material. Various techniques were applied to smooth and average errors, in order to arrive at ever more perfect screws.

Using this and similar machines, Maudslay went on to produce tools for the production of ordinary screws, as well as screws for all kinds of machines, including those requiring the utmost precision for measurement. In his works, he introduced a bench comparator, capable of measuring lengths to a tenth of a thousandth of an inch. It was known as "The Lord Chancellor," as it was regarded as the final court of appeal in matters of accuracy.

It was a particularly well made screw that Maudslay had displayed in the window of his first workshop that secured him the contract to build Marc Brunel's block-making machinery (*see* **Portsmouth Block-Making Machinery**). An amateur mechanic, M. de Bacquancourt, admiring the screw in the window, went in to the shop to find out how it had been made; the visit made a lasting impression on him. When de Bacquancourt heard that his acquaintance Brunel needed a good workman to implement his mechanical ideas, he made the introduction that made Maudslay's career.

During the period from 1800 to 1810, Mr. Maudslay effected nearly the entire change from the old, imperfect, and accidental practice of screw-making…to the modern, exact, and systematic mode now generally followed by engineers. —Charles Holtzapffel, 1846

—MICHAEL WRIGHT

HERSCHEL'S PRISM AND MIRROR (c. 1800)

At first glance, these objects, Herschel's prism and mirror, look insignificant. One is a small prism of flint glass about five inches long, the other an ellipsoidal metal mirror attached to a square frame. Their significance lies in the fact that they were used by William Herschel (*see* **Herschel's Seven-Foot Telescope**) in the first investigations of the spectrum beyond the visible region, now known as the "infrared." This pioneering work is probably less well known today than is Herschel's discovery of the planet Uranus or his construction of large telescopes, but the implications of extending the spectrum were far-reaching.

In 1799, during the course of his research on sunspots, Herschel (1738–1822) on many occasions viewed the sun through a telescope fitted with protective lenses. To cut out the glare, he used densely colored glasses in the eyepiece. This solution was, however, beset by problems: if the glare was satisfactorily reduced, an intolerable heat was transmitted: if the heat was reduced, the lens transmitted too much light. Sometimes the glass cracked, on one occasion "with a very disagreeable explosion that endangered the eye." Having tried many colors, he was led to question whether, in his words, "the power of heating and illuminating objects might not be equally distributed among the variously colored rays."

His first experiments, published by the Royal Society in 1800, compared the heating and lighting effects of the various colors. Using this small prism, Herschel made a spectrum of sunlight fall onto a table where he then placed three thermometers. He used a piece of card with a hole in it to select each color in turn, the colors being those defined by Newton: red, orange, yellow, green, blue, indigo and violet. One thermometer was placed in the colored ray, while the others were used as controls, simply measuring the temperature of the room. Herschel found that the red gave considerably more heat than did the other colors, the quantity decreasing to almost nothing for the violet ray. He then used the prism in a second experiment, viewing objects under a microscope in different colored light. Judg-ing by his own eyes, he found green/yellow to be the best for illumination.

Herschel found that dark-green smoked glass provided him with the best results for his telescopic research, but, more interestingly, he demonstrated that the maximum heating effect and lighting effect are at different points on the spectrum. A more tantalizing prospect was that the maximum heating effect might lie "even a little beyond" the red end of the visible range. Herschel set out to investigate this possibility.

Again the prism was used to create a spectrum, but this time, one of the thermometers was placed beyond the red end. A card was marked in inches, so that the position of the maximum heat could be found. As he expected, Herschel found the great-est heating effect just beyond the red, but he found no effect beyond the violet. This was the first time that experiments had been made on the spectrum outside the visible region.

Herschel went on to perform many experiments on "the rays which occasion heat." He used his

Herschel's glass prism on a brass base, used for his experiments on thermal radiation

kitchen fire, a candle flame, and a red-hot poker. He showed that heat could be reflected, refracted, or bent, scattered and transmitted in much the same way as light. The mirror was used for experiments on the solar spectrum. He attached the square frame holding the mirror to a wall or shutter, and he turned the mirror slowly to keep the sun's rays reflected on to the prism.

Like nearly all of his contemporaries, Herschel believed that light consisted of a stream of parti-cles. In his description of the first experiment, he surmised that "radiant heat consists of particles of light of a certain momenta." Paradoxically, in the same year, Thomas Young (1773–1829) proposed his wave theory of light. Although this theory was not well received, Herschel had to take account of Young's work, and he referred later in the year to "vibrations which occasion heat or rays."

Although an explanation of Herschel's findings based on the wave theory of light had to wait until the 1830's, there was a faster response to the idea of infrared from Johann Wilhelm Ritter (1776–1810) of Silesia (now Poland). Because he believed in the harmony of nature, Ritter looked for an effect at the violet end of the visible spectrum. In 1801, he found that these rays turned paper soaked in silver-chloride solution black, and hence he was the first to observe "ultraviolet" radiation.

We now believe that the "infrared," "vis-ible" and "ultraviolet" regions are a small part of an entire electromagnetic spectrum that includes radio waves, microwaves, x-rays and gamma rays. All of these kinds of radiation are simultaneously particles and waves, and they travel at the speed of light. Obviously, a great deal of thought and experi-ment has been applied to this area of physics since Herschel's day, yet he took the first step with this simple apparatus.

—JANE WESS

COALBROOKDALE BY NIGHT (1801)

… if an atheist who never heard of Coalbrookdale could be transported there in a dream, and left to awake at the mouth of one of those furnaces, surrounded on all sides by such a number of infernal objects, though he had been all his life the most profligate unbeliever that ever added blasphemy to incredulity, he would infallibly tremble at the last judgment that in imagination would appear to him.

With those apocalyptic words, Charles Dibdin, dramatist and songwriter, concluded his observations on the Severn Gorge in Shropshire, England. What drew Dibdin and numerous other travelers there in the late 18th and early 19th centuries was the remarkable concentration of industrial activity, concerned mainly with ironmaking, spread along some three miles of the valley of the River Severn between Coalbrookdale and Coalport.

The pioneering work of the Quaker ironmaster Abraham Darby in developing coke smelting in 1709 had been followed by a sequence of new uses for iron—the first iron rails, the first iron bridge, iron boats and steam-railway locomotives—making Coalbrookdale an area irresistible to anyone interested in understanding the new forces of industry that were sweeping Britain. Thomas Telford, William Jessop, Josiah Wedgwood, Richard Trevithick, James Watt, Matthew Boulton and John Loudon McAdam were the most notable of the British engineers and entrepreneurs who came to Coalbrookdale. They were joined by visitors from France, Prussia, Bavaria, Sweden and North America, many of whom were to leave accounts of what they saw. At a time when almost all tourists kept journals of their travels, many of which were published, these reflections provide a vivid and often technically detailed record of an area that the Shrewsbury cotton master, Charles Hulbert, was to describe in 1837 as "… the most extraordinary district in the world."

The gorge exercised a peculiar fascination for artists too. Here, the combination of topography— the River Severn flowing through its precipitous gorge—and the smoke and flames of the furnaces and forges conspired to produce a scene at once picturesque and sublime. Thomas Rowlandson, Paul Sandby Munn, John Sell Cotman, Joseph Farington and J. M. W. Turner are perhaps the best known of the numerous artists who visited the area during this period. But one work above all others epitomizes the artist's vision of the new industrial Britain. It has become the archetypal visual description of the first industrial nation.

Coalbrookdale by Night, painted in 1801 and pictured here, is the work of Philippe Jacques de Loutherbourg (1740–1812). Born in Strasbourg, the son of a painter of miniatures who moved the family to Paris when Philippe was still a boy, he settled in England in 1771 and was to have a profound influence on the new generation of artists at the end of the 18th century. His early reputation as a designer and painter of stage sets at the Drury Lane Theatre in London was further enhanced in 1781 when he opened the "Eidophusikon" in Lisle Street, Leicester Square, a precursor of the popular Panorama and a very real ancestor of the modern cinema. Dramatic effects were achieved by manipulating colored glass in front of the oil lamps that illuminated the scenery, while clouds painted on translucent linen moved diagonally past.

The powerful visual drama of the new industries had a special fascination for de Loutherbourg. He was drawn inevitably to Coalbrookdale. In *Coalbrookdale by Night* the buildings of an ironworks are silhouetted against the glare from the pig bed as the furnace is tapped. In the foreground, horses pull a wagon on a plateway among great castings. The precise setting for the picture was for long uncertain, but comparisons with contemporary views, notably a watercolor of 1803 by Paul Sandby Munn, demonstrate beyond all doubt that the scene is of the Bedlam Furnaces. Although Bedlam Furnaces are downstream of the Iron Bridge, at the time of de Loutherbourg's visit the whole area of what is now called the Ironbridge Gorge was still known generically as Coalbrookdale. Today, the name is restricted to the small side valley of the Severn where Darby established his ironmaking enterprise and where the Coalbrookdale Company still makes iron castings.

—Neil Cossons

PORTSMOUTH BLOCK-MAKING MACHINERY (c. 1803)

In the early history of mass production, Portsmouth block-making machinery, for making ships' pulley blocks, designed by Marc Isambard Brunel (1769–1849) and built by Henry Maudslay (1771–1831), takes pride of place.

Earlier attempts at the mechanization of production were mostly limited to the making of rather simple goods, such as "cards" for preparing cotton fibers for spinning. The usual approach to the production of an article in quantity was to subdivide the work into a large number of small manual operations, which were then performed at amazing speed by workers, each of whom specialized in one activity.

In 1800 (when Britain was at war with France), the Royal Navy's requirement for ships' blocks was very great—about 100,000 a year. It is not surprising that the Royal Navy's larger suppliers had already introduced a considerable degree of mechanization. Blocks are simple enough in themselves, but their manufacture involves more than twenty separate operations. What was new about Brunel's proposal was the concept of a series of machines performing those operations with no need for skilled labor, leaving to human hands the moving of pieces and the placing them in the machines, the working of levers, and the final assembly. Brunel devised special-purpose machines, the number of each type chosen to ensure a steady flow of components through the several operations from raw materials to assembly.

Brunel had been a French naval officer; his royalist sympathies caused him to flee to America following the French Revolution. There he practiced as an engineer, advancing to the post of Chief Engineer of New York. He came to England in 1799 to marry Miss Sophia Kingdom, whom he had met in France. While he was in America, a chance conversation had turned his thoughts to the problem of the manufacture of blocks by machinery, and by 1801, his ideas were far enough advanced for him to apply for a patent.

Through his brother-in-law, who was Under Secretary to the Navy Board, Brunel first offered the use of his invention to Messrs. Taylor of Southampton, the Navy's principal supplier of blocks. Samuel Taylor's response was at once complacent and harassed; he suggested that the firm's capacity was so stretched that they could not afford the time to consider "improvements." Brunel was then introduced to Sir Samuel Bentham (1757–1831), Inspector General of Naval Works, who had himself studied the application of machinery to woodworking. Because he appreciated the merit of Brunel's scheme, Bentham persuaded the Navy to set up its own factory at Portsmouth Dockyard to implement that scheme.

Brunel's machinery called for superior workmanship, and in the first instance he required

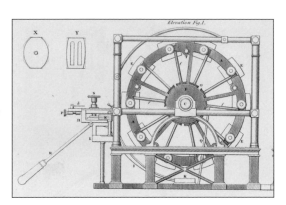

The shaping engine: an engraving from Rees' Cyclopaedia, *1819*

working models. Another French émigré, M. de Bacquancourt, had already recognized Maudslay's exceptional ability and had made the appropriate recommendation (*see* **Maudslay's Screw-Cutting Lathe**). Brunel first showed Maudslay a drawing of one machine without disclosing its purpose. Then he showed him a drawing of another machine, at which Maudslay exclaimed, "Ah! Now I see what you are thinking of—you want machinery for making blocks." It should be mentioned that the ability to read a drawing was not then common among working men. Brunel had found the high-achieving mechanic that he needed, and Maudslay got the contract that launched his career.

Whether the final designs were wholly Brunel's or included a significant input from Maudslay has been the subject of much debate. Certainly, the block machinery was highly advanced, in that it comprised large machines built entirely of metal. This provided the sturdiness that allowed Maudslay to build in such a way as to achieve great precision of action. As a consequence, the machines ran well for one and a half centuries. Previously, all but the smallest machines consisted of a wooden framework, and Brunel's much publicized tools set a new standard for engineers to emulate. At the time, few other engineers could have taken such a step.

Brunel's new machines incorporated many radical innovations. The morticing machine worked automatically, and was the forerunner of both woodworking machines and slotting machines for metal cutting. The shaping engine rounded the outsides of ten blocks with a cutter moved by hand but held firmly in the machine and guided by templates.

—MICHAEL WRIGHT

TREVITHICK'S HIGH-PRESSURE ENGINE (c. 1806)

The steam engines of the 18th century, the engines of Newcomen and of Watt, were low-pressure machines. That meant that their power output was small in relation to their size. A typical small rotative engine by Watt (*see* **Boulton and Watt Rotative Engine**) had an output of about ten horsepower (roughly that of a small car), but its boiler was as big as a substantial family house. Moreover, it depended on a continuous supply of cold water for condensation.

In all of these engines, steam fills the cylinder and is then condensed by cooling, to cause a "partial vacuum," a drop in pressure, which gives rise to the working force. The idea of using a much greater steam pressure to push the piston, without the element of condensation, was not a new one: it was described by Jacob Leupold in his *Theatrum Machinarum Hydraulicarum* (1725) and again by Watt in his patent specification of 1769.

But to contain and use high pressure called for an advance in engineering skill—and an engineer of bold determination. Richard Trevithick (1771–1833) was just such a man. Brought up near the mines of Cornwall, England, he followed his father's occupation as an engineer. A big man who delighted in such feats as hurling the smith's sledgehammer over the engine house, he more than made up for a lack of formal schooling by his intuitive grasp of mechanics.

In his patent specification of 1802, Trevithick outlined both the simple high-pressure engine and some possible applications. As it was much smaller than earlier engines and did not need to be connected to a supply of cold water, Trevithick's engine was obviously appropriate for driving road carriages or railway locomotives. Trevithick began experimenting immediately with both kinds of machines.

The most typical application was, however, for stationary tasks such as pumping, winding, or driving machinery. Because it was self-contained and required only the simplest of foundations, Trevithick's engine was quick, convenient, and cheap to install, even for temporary work such as excava-

A Trevithick high-pressure engine used in a dredger: engraving from Rees' Cyclopaedia, *1819*

tion for buildings. For this reason, despite its great weight, it was often known as a portable engine. If users noticed that it did not quite live up to the exaggerated claims in its advertising—that it was notable for economy of fuel usage—that did not prevent its widespread use; nor, in the long run, did the disastrous explosion of an early machine (through gross mismanagement) at Woolwich.

Trevithick's high-pressure engine (pictured opposite) is small—only three horsepower. Even so, it is much more compact than earlier versions were. The boiler shell—of cast iron, about forty millimeters thick—forms a framework to which the other parts are fixed. The cylinder is fitted vertically into the boiler, an arrangement that keeps it hot and reduces the steam passages to a minimum. Steam is distributed to either end of the cylinder, where it pushes the piston up and down, and from there to the exhaust by a simple plug cock. The exhaust steam passes through a feed-water heater before being released into the chimney. The fireplace is in the mouth of a tubular flue that runs forward and then back to the chimney.

Engines such as this one were shipped all over the world; Trevithick himself went to Peru to set up his machinery in silver mines there. Thus, it is not unreasonable to assume that there are more of his engines yet to be discovered in various parts of the world. But, apart from one other much less complete, the engine pictured opposite is the only known preserved example.

—Michael Wright

DALTON'S WOODEN ATOMS (c. 1810)

Atoms are round balls of wood invented by Dr. Dalton —answer given by a pupil to a question on atomic theory, as reported in 1887 by Sir Henry Enfield Roscoe to the British Association for the Advancement of Science.

The concept of "atoms," a microscopic entity beyond which matter cannot be divided, was known to the ancient Greeks, but it was not universally accepted. The famous natural philosopher Isaac Newton developed an atomic theory in which the chemical properties of substances (such as their acidity) were explained by their shape. Although his ideas were fairly popular, Newton's atoms did not become an accepted precept of chemistry at the time. The modern understanding of atoms began with a rustic Quaker from the Lake District of England named John Dalton (1766–1844), who was born in Eaglesfield, near Cockermouth, England. In 1793, he went south to make a living teaching mathematics and philosophy in the booming industrial town of Manchester. His original interest was in meteorology, and he consequently began to study the properties of gases. Dalton soon realized that he could explain some of these properties by assuming that gases were made up of tiny hard spheres. He then came up with the crucial idea that these spheres had different weights; this is our now familiar concept of atomic weight. He presented his theory to the Manchester Literary and Philosophical Society in 1803, and within a few years he had created a set of symbols (not related to our current system) to represent the most common elements.

Using these symbols, Dalton speculated as to how atoms were combined in compounds, even such complex molecules as ether and ethyl alcohol. He made assumptions of how atoms combined, and assigned what he felt was the logical structure for any given compound—for example, "OH" was his formula for water. He was not a chemist, however, and he did not know how to prove experimentally his ideas about molecular structure.

Despite these difficulties, Dalton made an important heuristic advance, one that was to change radically the very concept of atoms. He had the idea of making physical representations of atoms, and in about 1810, he convinced his friend Peter Ewart, a millwright and engineer, to make a set of wooden atoms. They were smooth wooden balls, about a centimeter in diameter, that were connected by small wooden dowels. Dalton used them in his lectures, probably to show the differences between gases, liquids and solids in his atomic system; their main use was to explain physical, rather than chemical properties.

Yet Dalton's atoms appear to have highly advanced features. Although it is not clear in photographs of the atoms today, traces of color can be seen on the surface. It is quite possible that Dalton used different colors for different elements, thus anticipating August Wilhelm von Hofmann's use of table-croquet balls in 1862 to represent the elements. Furthermore, the holes for the dowels are arranged all over the surface of the atoms. His molecular diagrams certainly show that he had considered the arrangement of atoms in two dimensions. Simply by using wooden balls to represent atoms, Dalton was open to the possibilities of the arrangement of atoms in space. Were the holes just for convenience, or did Dalton speculate about the arrangement of atoms in space? If he did so, he was far in advance of other chemists of his day.

Similar ball-and-stick kits did not appear until the 1860's, and the arrangement of atoms in space had to wait until the work of Jacobus Henricus van 't Hoff and Joseph-Achille Le Bel in the mid-1870's. By embodying his concept in a tangible form, Dalton made atomic theory more comprehensible to the scientist and the layperson alike. Although it took more than a half-century for Dalton's theory to be accepted by most chemists, the concept of atoms has become central to modern science.

Since the discovery of radioactivity in 1896, atoms are no longer regarded as indivisible hard spheres, and elements are no longer thought to include atoms of a single weight—but rather mixtures of isotopes of different weights. Like Newton's theory of gravity, Dalton's atomic theory is considered to be fundamental; it has been largely discredited in terms of its details. Perhaps Dalton's enduring legacy to chemistry today is the ingenious notion of using balls and sticks to represent the molecular structure of compounds. For amateur chemists trying to work out the structure of a compound, atoms are indeed "round balls of wood invented by Dr. Dalton."

—PETER MORRIS

THE *COMET* STEAM ENGINE (1812)

The earliest steam engines were too heavy to be installed in ships, but following the improvements made by James Watt and Richard Trevithick (*see* **Boulton and Watt Rotative Engine** and **Trevithick's High-Pressure Engine**), steam engines were employed in both Continental Europe and North America to propel ferries and to tow barges on lakes and inland waterways. These early services were highly experimental, and accidents involving boilers caused many setbacks. As well, Scotland's Forth and Clyde Canal was the scene for trials of the paddle tug *Charlotte Dundas* in 1801, but damage from her wash is alleged to have annoyed other canal users, and she was withdrawn from service. Although not always successes, such experiments were important in demonstrating that it was no longer necessary for ships to be completely dependent on favorable wind or tide.

In 1810, passenger service on the River Clyde below Glasgow was principally provided by "flyboats," fast rowboats the schedules of which were necessarily governed by the strong tides on the river. Henry Bell (1767–1830) was a hotelier in Helensburgh near the mouth of the Clyde, and he envisioned a steamboat that could bring visitors to his hotel from Glasgow, whether or not the tide was favorable on the Clyde.

A wooden ship some forty-three feet long was built to Bell's specifications at John Wood's shipyard at Port Glasgow in 1811. Based on a typical sailing packet of the day, she was named *Comet* to commemorate a comet seen in Scotland that year. Steam-engine design at that time was directed principally toward stationary applications, and the engine that Bell ordered for his vessel might equally have powered a small workshop. This engine was created by John Robertson (1782–1868). When it was installed in *Comet*'s wooden hull, the engine's weight was balanced by that of a brick-mounted boiler on the starboard side, with a tall funnel between them doubling as a mast. Launched on

July 24, 1812, Bell's twenty-five ton steamer commenced a service between Glasgow and Greenock on Tuesdays, Thursdays and Saturdays, returning on the intervening days "to suit the tide;" evidently, she was insufficiently powered to oppose the tide. Replacement of her original eleven-and-a-half-inch diameter steam cylinder with one an inch larger helped to increase the power.

Despite opposition from the flyboat owners and ridicule from some skeptics, *Comet* continued her service on the Clyde, and within two months of her launch, several rival steamers were being built. Hoping that increased capacity would help *Comet* pay her way, Henry Bell lengthened her hull by twenty feet, retaining the engine provided by John

John Robertson with Comet's *engine; its original cylinder is in the left foreground*

Robertson. After a period of service on the Firth of Forth, *Comet* was used to establish passenger service between Glasgow and Fort William in the Scottish Highlands. Although the Hebridean Islands provided shelter for part of the route, Bell must soon have realized that *Comet* was grossly underpowered for such a service. He made up his mind to order a more powerful vessel, and on December 15, 1820, he was returning aboard *Comet* from Fort William to raise capital for it in Glasgow when a combination of wind and tide forced *Comet* ashore on Craignish Point. Bell and the other passengers scrambled off. *Comet*, weakened by its recent lengthening, broke in half.

Henry Bell still owed money to various people in Glasgow and Helensburgh. One creditor who was more determined than most was Bailie McLellan, a Glasgow coachbuilder and founder of the city's great art collection. He took a cart up to Craignish Point, recovered *Comet*'s engine from the rocks where it had lodged, and put it to work driving machinery in his Glasgow coachworks. In 1836, the engine was moved to another Glasgow factory, where its historical significance as Europe's first successful marine steam engine was recognized. It was exhibited at the Glasgow Polytechnic, but when that building caught fire in 1855, *Comet*'s engine fell from the top story and was buried in the debris.

The Glasgow engine-builder John Napier eventually succeeded in getting custody of the engine, and it was cleaned and reassembled with a respect for authenticity rare at that time. Napier's firm also took the trouble to track down John Robertson, by then more than eighty years old, and Robertson posed for photographs of himself standing beside the engine in the Napier works.

—JOHN ROBINSON

PUFFING BILLY (1813)

In both name and appearance, *Puffing Billy* is a classic example of an ancient mechanical relic. It is also one of the most important inventions from the earliest days of railroads: the first steam locomotive to operate successfully on a commercial basis, involving the adhesion of a smooth wheel to a smooth rail. Together with sister locomotive *Wylam Dilly*, preserved in the Royal Museum of Scotland, in Edinburgh, it is the oldest surviving steam locomotive.

To appreciate the significance of *Puffing Billy* in railroad history, one must consider the development of early steam locomotives. Richard Trevithick, an engineer from Cornwall, England, was the first to adapt the stationary steam engine into a machine capable of propelling itself along a self-guiding track while hauling wagons (*see* **Trevithick's High-Pressure Engine**). His first design, built at Coalbrookdale in 1802, was not successful, but his second, built in 1804, demonstrated at Pen-y-darren in South Wales the potential of a steam locomotive to replace horses for hauling coal wagons. The boiler on these early designs was comparatively large and heavy. Although this weight helped with adhesion, it also was the cause of many broken rails.

News of the success of these limited trials, and their potential significance for the coal industry in particular, soon spread. Christopher Blackett (1751–1829), the owner of Wylam Colliery in Northumberland, was interested in the potential of the Trevithick designs to replace horses on the five-mile-long wooden wagonway between his coal mine and the head of navigation on the River Tyne. Blackett was deterred from this goal, however, by concerns that the weight of a locomotive might damage the wooden rails.

Early in 1812, Blackett's interest was revived after a more advanced locomotive, one with two vertical cylinders, operated successfully on a rack railway at Middleton Colliery near Leeds. Here, a toothed wheel on the locomotive engaged similar teeth on a continuous rail, thus enabling the train to overcome steep gradients. There were still, however, difficulties and limitations at junctions in the track.

The quickening pace of industrial growth placed a premium on productivity, and this imperative, together with the increasing price of horse fodder during the Napoleonic Wars, further encouraged entrepreneurs to replace the horse with more efficient motive power. The relatively level Wylam Wagonway had been relaid in 1808 with a cast-iron plateway, to the same five-foot gauge as the previous wooden rails. Rather than incur the cost of relaying again with the more expensive and limited rack rails, Blackett decided to build a steam locomotive that could operate on smooth iron rails. The challenge was to design one that would provide sufficient adhesion to pull a commercial load on a conventional railway without an unacceptably high rail-breakage rate.

In October 1821, Blackett asked his coal mine manager, William Hedley (1779–1843), to build such a locomotive. It was constructed during 1813, probably in Gateshead, with a single cylinder and a flywheel. It failed in trials primarily because it didn't produce enough steam, having only a single heating flue in the boiler. Another locomotive, with two cylinders and a return flue, was built, probably at Wylam, sometime during the winter of 1813–14. This locomotive probably entered service in March 1814; it could successfully haul up to nine or ten coal wagons at speeds of four to five miles per hour on the Wylam Wagonway. This is the locomotive that became known as *Puffing Billy*.

It was followed later in 1814 by two more locomotives, subsequently known as *Wylam Dilly* and *Princess Mary*. Contemporary reports mention the strange noise made by the engines, and, indeed, there is evidence that on *Puffing Billy,* the exhaust steam passed through a chamber on top of the boiler designed to reduce this noise. The name *Puffing Billy* is actually a nickname: "Billy" being a corruption of the word "dilly," which was a local name for the caldron, or coal wagon, used on the Wylam Wagonway.

Although the locomotive as built was successful, there were still problems with damage to the plateway. Contemporary sketches show that the original four-wheel locomotive was modified to an eight-wheel in order to spread the weight. At some later stage, perhaps when the plateway was rebuilt in about 1828 (with stronger cast-iron edge rails, but still to the five-foot gauge), the locomotive reverted to its original four-wheel configuration, which it then retained. The flanged wheels that it now possesses would no doubt have been added at this time. In other respects, the locomotive appears very much as built, although comparison with early drawings indicates that at some stage the valve gear was modified.

Puffing Billy continued working at Wylam until the early 1860's. The modest task for which the locomotive was intended meant that its performance was adequate for many years. That fate is in contrast to many of its illustrious successors such as *Rocket* (*see* **Stephenson's Rocket**); the pace of development of locomotives at the forefront of railway evolution led to a much shorter working life for these locomotives, at least for the work for which they were originally built.

Despite the success of *Puffing Billy*, much remained to be done before the full potential of the railway could be realized. Perhaps the greatest underlying significance of *Puffing Billy* is that its success encouraged George Stephenson to pursue his interest in steam locomotives and railways. It was he who was to develop the steam locomotive to the point at which the success of the railways, as we know them today, was finally proven.

—JOHN COILEY

DAVY'S SAFETY LAMP (1815)

Davy's safety lamp, designed by Sir Humphry Davy (1778–1829), is said to be the first safety lamp actually used in a coal mine.

In the 18th century, miners worked by the light of candles or other naked flames. As mines became deeper, they encountered a new hazard: firedamp. It is a gas, composed mainly of methane, produced by the decomposition of vegetable matter when coal seams are formed; it is retained in fissures and molecular interstices. The gas is not poisonous, but when it is mixed with air in certain proportions, it is explosive, and it can be ignited by a naked flame.

An increasing number of miners lost their lives in pit explosions caused by firedamp, and several attempts were made to develop a lamp that would be safe underground. Carlisle Spedding (c. 1696–1755) invented a steel-and-flint mill: light was created in the form of a stream of sparks caused by rotating a thin steel disk against flint. This invention was widely adopted in the north of England because it was safer than a candle. Explosions still occurred, however, and boys were usually needed to accompany miners to operate the device, so it was a costly solution.

On May 28, 1812, an explosion at Felling Colliery killed ninety-two men and boys. The publicity given to this tragedy by the vicar of the parish, the Reverend John Hodgson, resulted in the establishment, on October 1, 1813, of the Sunderland Society for Preventing Accidents in Coal Mines. The Society approached Sir Humphry Davy for help. Davy met with John Hodgson and John Buddle, the leading mining engineer of the day, and he visited Hebburn Colliery, where he carried out experiments. After Davy returned to London, Hodgson sent him some firedamp in six wine bottles. Shortly afterward, Davy wrote: "I have already discovered that explosive mixtures of mine damp will not pass through small apertures or tubes; and that if a lamp, or lanthorn be made air tight on the sides, and fur-

Sir Humphry Davy

nished with apertures to admit the air, it will not communicate flame to the outward atmosphere."

Subsequently, on November 9th, Davy read a paper to the Royal Society entitled, "On the firedamp of coal mines and on methods of lighting the mines so as to prevent its explosion." Three lamps were described in the paper. They were glass lanterns, with an air feed through concentric metal cylinders or, in one case, through wire gauze. The tops of the glass chimneys were similarly equipped. By this time, Davy had discovered the principle on which the safety lamp was to be constructed, but it was another month or so before he conceived of the idea of surrounding the flame of the lamp with wire gauze.

Davy's first safety lamps were completed at the end of 1815 or in early 1816 and were sent to John Buddle. They were tested in Hebburn Colliery on January 9th and 17th. John Buddle recorded his reactions: "To my astonishment and delight, it is impossible for me to express my feelings at the time when I first suspended the lamp in the mine, and saw it red hot; if it had been a monster destroyed I could not have felt more exultation than I did. I said to those around me, 'We have at last subdued this monster.'"

Each lamp has a cylinder of brass wire gauze of one and a half inches in diameter and five inches high, screwed to the top of a reservoir of oil. In one, there is also a spout for filling the vessel and a pricker to trim the wick.

Davy was well aware that, under certain circumstances, such as a strong current of firedamp, his lamp could be unsafe, and he suggested some modifications. They included a gauze cap over the top of the main gauze, to cool the burnt gases, and a tin shield sliding on the frame wires of the lamp, to protect it from drafts.

The Davy lamp was widely adopted. Over the years, many modifications and improvements were made. Davy was awarded a service of silver plate worth £2,500 as compensation for his efforts. Unfortunately, the Davy lamp did not reduce the number of accidents and fatalities; it merely allowed for the exploitation of deeper and more dangerous mines.

At about the same time as Davy was engaged in his experiments, George Stephenson (1781–1848) was working on a safety lamp that was also based on the principle of explosive gases not passing through small tubes and apertures. There arose considerable controversy, never fully resolved, about which man had created the first safety lamp. In practice, Davy's lamp was far more significant because it was his understanding of the scientific principles and application of the wire gauze that was so valuable and formed the basis for the design of virtually all flame-safety lamps thereafter. Indeed, Stephenson also soon adopted wire gauze for his own lamps.

—JANE KIRK

FARADAY'S CHEMICAL CHEST (1825)

In the 1820's, the London Portable Gas Company (which employed Michael Faraday's brother Robert) provided containers of gas for use in lighting houses, in much the same way that today propane gas containers are used for cooking and heating. The gas in question was not, however, a by-product of petroleum refining, but was made by heating whale oil, then a common article of commerce. A puzzling liquid sometimes condensed out of this gas, and in 1825 the company sent a sample to Michael Faraday (1791–1867) for analysis. By the process of fractional distillation, Faraday isolated a large fraction that distilled at 176°F to 190°F (80°C to 88°C). Modern analysis has shown that a sample of Faraday's benzene is remarkably pure: 99.5%. To achieve this level of purity, Faraday must have cooled the benzene to its freezing point of 5.5°C, then used fractional crystallization to purify it—a remarkable demonstration of his experimental skill. He then determined its chemical formula as C_2H. Because he assumed that the atomic weight of carbon was six, he had in fact found the correct empirical formula, CH; he called it carbureted hydrogen.

In the work of later chemists—notably Eilhard Mitscherlich, who coined the name "benzin," and August Wilhelm von Hofmann—benzene became the central compound of aromatic organic chemistry. Its striking chemical stability was first explained in the mid-1860's by August Kekulé, who argued that benzene was a ring of alternating double and single bonds, a configuration that gave the molecule its stability. Chemists have continued to explore this phenomenon of "aromaticity," and have found other compounds that have a similar stability. The modern technique of nuclear magnetic resonance (NMR) has shown that benzene and other aromatic compounds have a tiny electric current in their rings. It is thus appropriate that Faraday, one of the "fathers of electricity," discovered a compound with its own inherent electric current.

The status of benzene as a major industrial chemical began in the late 1840's when Charles B. Mansfield, a former student of Hofmann at the Royal College of Chemistry (now part of Imperial College, London) distilled benzene from coal-tar. Benzene was found to be an excellent solvent, and despite Mansfield's death in 1855 from burns caused by his distillation apparatus, the flammability of benzene was tolerated. An even more momentous step was the accidental discovery in 1856 of the first important synthetic dye, mauve, by another Hofmann student, William Henry Perkin. Ironically, Perkin was able to make mauve only because commercially available benzene was impure, compared to Faraday's pure product. Benzene soon became the basis of numerous dyes and pharmaceutical products.

Faraday's mentor, Humphry Davy, had shown in 1810 that hydrochloric acid (contrary to the theory of Antoine Lavoisier) contained no oxygen. Ten years later, Faraday showed that carbon could combine with chlorine to form what he called carbon perchloride—when ethylene and chlorine gas were exposed to sunlight. To Faraday's surprise, this perchloride of carbon (now called hexachloroethane) was chemically stable, indifferent to the action of silver nitrate or sulfuric acid. During this research, Faraday also isolated another carbon-chlorine compound that he called carbon protochloride (we now call it tetrachloroethene or perchloroethylene). This compound was also surprisingly stable. Faraday did not pursue this line of research, but a decade later two French chemists, Jean Baptiste Dumas and Auguste Laurent, substituted chlorine for hydrogen in organic compounds, which led to a complete reassessment of the formation of organic compounds and, in turn, to the establishment of modern organic chemistry.

For many years, Faraday's carbon chlorides were a laboratory curiosity until the early 1900's, when cheap chlorine became available as a by-product of electrolytic caustic soda production. The Bavarian firm, Wacker-Chemie GmbH, and its British associate, Weston Chemicals, developed the use of chlorinated hydrocarbons, including Faraday's perchloroethylene, as solvents. Gradually beating off competition from the closely related tetrachloroethane and trichloroethylene, perchloroethylene has become the important solvent in dry cleaning. It has the advantage of being less toxic than other chlorinated hydrocarbons, and, unlike its former high-tech rivals R-11 (trichlorofluoromethane) and R-113 (trichlorotrifluoroethane), it does not attack the ozone layer.

The Science Museum's glass tubes of benzene and hexachloroethane were found in the drawers of Faraday's chemical chest. This portable laboratory was donated to the museum in 1908.

—Peter Morris

LISTER'S MICROSCOPE (1826)

It has been suggested that Lister's microscope is possibly the most important optical microscope ever made. It was designed by Joseph Jackson Lister (1786–1869), father of the famous surgeon Joseph Lister, in March 1826, and was made by William Tulley, the London optical-instruments maker. It included improvements such as graduated draw tubes (to make it easier to set up and focus), lenses to act as a sub-stage condenser (increasing the amount of light going through the microscope slide), and a rotating and clamping stage to manipulate the slide. Its main significance is in the superb quality of the objective lens (the one nearest the specimen), from which distortions and color-change effects had been eliminated to a greater degree than ever before.

Joseph Jackson Lister's interest in optics can be traced back to his school days, and for the rest of his life, he designed and tested lenses and other microscopical equipment. Despite being a superior student in schools at Hitchin in Rochester and finally Compton in Somerset, Lister left school at the age of fourteen to join his father's wine business. His own talents and his father's generosity were demonstrated four years later, when in 1804 he became a full-fledged partner. In 1821, he became part owner of a ship commanded by his brother-in-law, and he continued with these and other business interests throughout his life, as adjuncts to his scientific and religious activities. He was an active and influential Quaker, and he became a school visitor and trustee.

It was as a result of his educational activities that he met his future wife, Isabella, who taught reading and writing until the day of their marriage.

Lister was elected to a fellowship of the Royal Society in 1832, on the strength of his work in microscopy and histology (the study of tissues). Of his seven children, only his son Joseph was to reach or surpass his position; his oldest son John died of a brain tumor in 1846 at the age of twenty-four. In those days it was quite unusual for all of the children in a family to reach adulthood, and Lister never lost sight of the wider world of suffering. In 1848, he not only served on a committee to raise funds for the relief of Irish potato famine victims, but as well he took the trouble to visit them to discover for himself the conditions under which they were living.

As a result of Lister's developments in lens design, the true appearance of tissues could be accurately described for the first time. Blood corpuscles were seen to be concave disks rather than globules, and muscle was shown to be made of fibers, with cross striations, regardless of the source of the muscle. In the spring of 1827, Dr. Thomas Hodgkin published jointly with Lister the results of these and many other observations on nerves, arteries, cellular membranes and the brain, refuting the views of some of the most eminent researchers of the time.

Lister published his further work on objective lenses in 1830, in the *Philosophical Transactions of the Royal Society*, and in doing so, he established for the first time the compound microscope (one with two main lenses) as a serious research instrument. He also continued to use his invention himself, and from the end of 1830, he was making his own lenses, despite having no previous experience in doing so. His own research interests extended into investigations of the limits of human vision, and he published another paper on zoophytes (microscopic animals), illustrated by his own drawings from the camera lucida, an attachment that allows the microscopist to view the specimen through the microscope, and, at the same time, "trace" outlines of that specimen directly on to the paper of a sketch pad.

Lister continued to take an interest in the design of the microscope, advising the London makers Andrew Ross, and later, James Smith on elements of lens design. With both of these men, Lister's associations were long and commercially invaluable, with some of Lister's improvements being widely adopted as standards.

Because of his naturally diffident nature, much of Lister's work was never published under his own name. The full extent of his many interests has only recently become known, as a result of a detailed study of his drawings, many of which were given to the Royal Microscopical Society on the death of his son Joseph. The true significance of his work can now be more widely appreciated.

—JANE INSLEY

BELL'S REAPER (1827)

Bell's reaper, which is now regarded as the first truly practical mechanical reaper, was designed in 1827 by Patrick Bell (1799–1869), later Reverend Bell, of Carmyllie, Scotland. It incorporated several features that are still seen in the modern binder, stripper and combine harvester. Yet Bell's pattern was never produced in large numbers, and it was to be the success of American designers—most notably Cyrus McCormick at the Great Exhibition of 1851—that would lead eventually to the establishment of mechanized reaping in Great Britain. Only then was the Reverend Bell given the opportunity to assert his importance in the history of the reaper's invention.

Reaping (the cutting of the corn at harvest) had for centuries been achieved using handheld sickles and scythes. A mechanical reaper is, however, described in the writings of ancient Roman scholars, and in 1787, a depiction of a machine based on these accounts appeared in the periodical *Annals of Agriculture*. The author/illustrator in the 1787 article made no attempt to build such a machine, but by the end of the 18th century, several mechanics had constructed reapers based on the Roman model. All of these reapers proved to be unsuccessful.

Some advances were made in the early years of the 19th century; for example, full-size machines were constructed by such pioneers as Thomas Plucknett, Joseph Mann, and James Smith of Deanston (1789–1850). Most of these machines had qualities that were ingenious, but they were all eventually abandoned. It was Bell's reaper that proved to be the first workable design.

Patrick Bell was the son of a farmer. He showed an early interest in mechanics, and for some years pondered ways to mechanize reaping in order, he later explained, to ease the burden of his father's workers at harvest time. Ignorant of previous research (except that of James Smith), he dismissed a succession of his own ideas until he began work on a design that incorporated the cutting action of garden shears. He made a small working model in 1827, followed by a full-scale prototype.

Bell was very secretive about his invention, and he went to extreme lengths to ensure that the early trials of 1828 were unobserved.

Toward the end of 1828, his reaper was successfully demonstrated at a small number of exhibitions. A specially commissioned reaper was also shown before the Highland Society of Scotland at Edinburgh, where Bell was awarded £50 (immediately paid to a creditor). Further exhibitions in 1829 were sufficiently successful that they created opportunities for Bell to construct a few machines for interested clients.

But any kind of wider interest was slow to develop, and by the mid-1830's fewer than twenty machines were in use, mainly in Scotland. As the reapers required regular maintenance, many of them gradually fell into disrepair when in the hands of inexperienced users. A further factor that led to their decline was Bell's reluctance to patent his ideas. His attitude resulted in varying standards of workmanship on his own reapers; it also allowed inferior copies to be made by other manufacturers.

International sales were negligible, but it has been suggested that several of his reapers were probably exported to North America in the early 1830's. This assumption is made because drawings of his reaper were published there. The degree of influence these drawings had on the American designers who were subsequently to dominate the market is an unresolved issue.

Bell was afforded widespread recognition belatedly after the Great Exhibition, where the American machines of McCormick and Obed Hussey had created a stir. Now powered by two horses, and with an improved cutting mechanism, Bell's reaper emerged from obscurity to compete against the American models in a series of trials that heralded the acceptance of the reaper in Great Britain.

Thus the real relevance of Bell's reaper was finally recognized, although it was not until 1867, only two years before his death, that Bell received a stipend of £1,000 from the Highland and Agricultural Society of Scotland in appreciation of his contribution to agriculture.

—STUART EMMENS

Bell's Improved Reaping Machine by Crosskill,
steel engraving by J.W. Lowry, after Cornelius Varley, c. 1840's

STEPHENSON'S *ROCKET* (1829)

In the history of the early development of the steam railway locomotive, there is no more significant machine than *Rocket*. Its design and construction proved to be a momentous leap forward in locomotive and railway systems, as it embodied the fundamental principles that were thereafter to be followed throughout the 150-year history of the reciprocating steam locomotive.

Stephenson's *Rocket* clearly demonstrated the superiority of moving locomotive engines over stationary engines and rope-haulage systems. Yet *Rocket* was an experimental machine at the forefront of a rapidly developing technology. It was soon modified, and it was rendered obsolete in little more than a decade.

The first steam locomotives were built for hauling loads rather than for speed (*see* **Puffing Billy**). They offered little advantage over stationary engines and rope haulage. The directors of the Liverpool & Manchester Railway (L&MR) wanted a faster and better form of motive power for their inter-city railway. They offered a prize of £500 for a "locomotive engine which shall be a decided improvement on those now in use." Competing engines in the competition were to be tested on a length of the new railway at Rainhill, east of Liverpool.

Rocket was built at the Robert Stephenson and Company locomotive works in Newcastle-upon-Tyne to compete for the prize at the Rainhill Trials, which were held in October of 1829. It was the eighteenth steam locomotive built by that company, and it relied on the experience of George Stephenson (1781–1848) and his son Robert (1803–59). It was Robert who undertook the construction of this innovative machine: it was a lightweight locomotive with a single pair of driving wheels, built for speed—as its name implies.

In the trials, *Rocket* was in competition with four other machines; it emerged the clear winner. The locomotive's success was due largely to a number of important design features that were incorporated by the Stephensons. The multi-tube fire tube boiler had been suggested to the Stephensons by

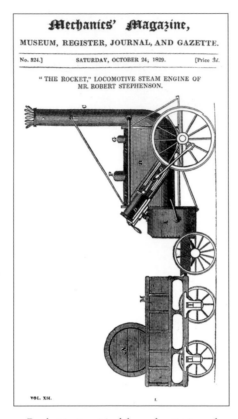

Rocket in its original form: frontispiece of Mechanics' Magazine, *October 24, 1829*

Henry Booth, and it was greatly superior to the more prevalent single- or twin-flue boiler. Twenty-five three-inch-diameter tubes in the boiler increased the heating surface, and, as a result, also increased the amount of steam that the boiler could produce. This enhancement was further assisted by the Stephensons' use of a water jacket around the firebox. Steam from the cylinders was exhausted through a blast-pipe up the chimney, which induced draft in the firebox and made the fire burn even hotter. *Rocket* was also built with steeply angled cylinders, which went directly to a single pair of driving wheels with crank pins at right angles. This was a more effective system than that of earlier engines, which had used vertical cylinders. The multi-tube

boiler, blast-pipe exhaust, and two-cylinder simple-drive mechanism are the three fundamentals of steam locomotive design: they were first combined in *Rocket*.

Although built expressly to compete for the prize (and not for practical, everyday use), *Rocket* so impressed the directors of the L&MR that they purchased it for use on the line and ordered four more similar locomotives from Stephenson. In celebratory ceremonies on September 15, 1830, *Rocket* was involved in an accident in which William Huskisson, then Member of Parliament for Liverpool, was killed. This was the first recorded passenger fatality on a railway.

Within eighteen months of its first construction, *Rocket* had been significantly rebuilt. The cylinders were swapped around and dropped to eight degrees from the horizontal to improve the riding of the locomotive, and the splayed-out chimney base was replaced by a more practical drum-shaped smokebox. Other changes included the addition of a buffer beam as well as a reduction in the chimney size.

By the mid-1830's, *Rocket*-type locomotives and their successors had displaced the prototype on the L&MR. *Rocket* was sold in 1836 to Messrs. Thompson of Kirkhouse, near Carlisle. It had a three-year working life on the Brampton Colliery Railway in Cumberland, but by about 1840 it was considered too lightweight and worn out to pull coal trains.

A proposal to show the locomotive at the Great Exhibition of 1851 came to nothing, and the derelict locomotive was presented by Messrs. Thompson to the Patent Office Museum in London in 1862. For public display, Stephenson and Company undertook some restoration work on *Rocket*. This restoration included the addition of parts incompatible with the rebuilt condition of the locomotive: some of these parts have since been removed. What remains are mainly those parts believed to date from 1829–36, the barebones of the locomotive in its rebuilt form.

—DIETER HOPKIN

BABBAGE'S CALCULATING ENGINES (1832)

Charles Babbage (1791–1871) was an English gentleman of science with a wide range of interests. He was an inventor, reformer, mathematician, philosopher, scientist, writer and critic. He is best known for his pioneering work on vast mechanical calculating engines, but Babbage also designed a variety of other devices: occulting lights used to communicate between ships; a camper van (with sleeping accommodation, cooking facilities, and a commode); speaking tubes for linking London and Liverpool; and even shoes for walking on water.

Before the widespread availability of electronic calculators in the 20th century, navigators, architects, engineers, mathematicians and bankers all relied on printed mathematical tables. These tables were produced through repetitive calculations performed by human clerks (who were literally called "calculators"). The results were then copied and set in loose type for printing. Not surprisingly, they were riddled with errors. Charles Babbage became interested in mechanizing the production and printing of the tables to eliminate these errors, and he developed a series of engineering diagrams and prototype machines that enabled him to explore his various solutions.

The task of mechanizing calculation had challenged many renowned intellects before Babbage. Devices were built in the 17th century by Blaise Pascal, the French philosopher-mathematician, and by Gottfried Leibniz. Pascal's invention was paraded before the social and intellectual elite of the day, and it caused something of a stir. These early desktop devices represent significant steps in the evolution of mechanical calculators, but they were more in the nature of ornate curiosities than serious devices for rapid and accurate calculation.

Babbage designed two kinds of engine: Difference and Analytical.

The Difference Engines are calculators that work on the mathematical principle of "a method of finite differences," which allows complex mathematical equations to be calculated using pure addition. Difference Engines are capable only of a fixed set of operations: today, we tend to refer to them as calculators.

Babbage's work on Difference Engine No. 1 started in the early 1820's. He worked with the engineer Joseph Clement to assemble a demonstration piece, which was just a small part of what was to be the completed 1832 machine. After a dispute with Clement in 1833, he abandoned his goal of creating a complete machine. The demonstration piece is, however, one of the most celebrated icons in the prehistory of computing; it is also one of the finest examples of precision engineering of its day. It is one of the first known automatic calculators, in the sense that mathematical rules were successfully incorporated into the mechanism. The operator did not need to understand the mechanical or logical principles in order to achieve useful results. All he had to do was turn the handle—the machine did the rest.

In contrast to his Difference Engines, Babbage's Analytical Engines mark a progression toward a "general purpose machine." The engines offered a repertoire of basic operations (multiplication, division, addition and subtraction), and they could follow instructions to perform a variety of different functions. The Analytical Engines were programmable, using punched cards—a technique already used in the Jacquard loom to control patterns woven with thread. The design of these machines included all of the essential logical features of the modern general-purpose computer, including a separate mill (the processor) and store (memory). The machines were even capable of "looping" (repeating the same sequence of operations a specifiable number of times) and "conditional branching" (choosing alternative actions in the form of "if… then…"). If an entire Analytical Engine had ever been completed in the 1830's, it would have been the size of a room and it would have been necessary to power it by steam.

This conceptual leap—from a machine that could calculate to a machine that was programmable—is one of the startling intellectual feats of the 19th century, and it accounts for Babbage's reputation today as an important computer pioneer.

Despite designing all of these machines, Charles Babbage never completed a whole engine. It was not until 1991 that the calculating section of Babbage's Difference Engine No. 2 was finalized. The completed engine has more than 4,000 parts, weighs five tons, and can calculate numbers up to thirty-one digits long. Two hundred years after Babbage's birth, it became clear that, had Babbage ever completed the machine, it would have worked. The most likely reasons for his failure were financial and political—and perhaps also due to his own contrarian personality.

As a result of the completion of the Difference Engine No. 2, we have begun to do justice to the grandeur and scope of Babbage's schemes, but it is the small section of the Analytical Engine, the "mill," that is the earliest precursor to the modern computer. Under construction at the time of Babbage's death, this model portion of the mill (from 1871) is a monument to humanity's earliest attempts to create an automatic general-purpose computer.

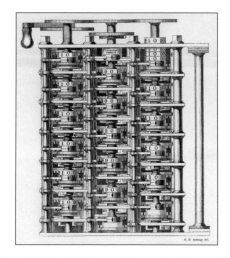

Above: Portion of Difference Engine No. 1 assembled in 1832; woodcut, 1853. Opposite: Experimental model of a portion of the Analytical Engine mill, 1871

—Tilly Blyth

What is generally regarded as the earliest existing photographic negative only measures 28 x 36 millimeters. It is mounted on a wafer-thin piece of black paper, which in turn is mounted onto a sheet of writing paper, on which are written the words:

Latticed Window (with the Camera Obscura)

August 1835

When first made, the squares of glass about 200 in number could be counted, with help of a lens.

This label, the mount, and the diminutive negative are the work of William Henry Fox Talbot (1800–77) of Lacock Abbey, Wiltshire. Talbot, though he was a wealthy man, was not satisfied to follow the role model established by those English gentry who managed their estates and generally led lives of pleasure. Instead, Talbot pursued ideas, using the patronage provided by his inheritance.

Talbot's "idea" for photography came from disappointment at his inability to draw Italian scenery using the "camera lucida." This was a portable instrument that used a small prism to project an image of a scene onto a piece of paper. He began to consider the agencies of optics, light and chemistry in an attempt to capture the elusive view.

The essential elements had long been available. Chemists in the 18th century knew that the salts of silver darkened with exposure to light. The "camera obscura" was made by opticians of the 17th century for use by artists interested in achieving correct perspective. (The term "camera obscura" means "darkened room:" it was a light-tight box in which a small hole or lens allowed light to enter so that an image of the view outside was projected on the side opposite the hole/lens.) It required Talbot's fertile imagination to bring these elements together and realize the potential for making a permanent record of the kind of image produced by the camera obscura.

In the spring of 1834, Talbot began his first experiments at Lacock. Some believe that Talbot had his estate carpenter make his first cameras.

In Talbot's system, a sheet of writing paper was brushed first with a solution of common salt. A second wash of silver nitrate combined with the salt to form light-sensitive silver chloride. The prepared paper was then gummed inside the back of the camera, where it was exposed until it darkened—for about an hour. Once the image (in which light and dark areas were transposed) was visible, the paper was removed from the camera and stabilized against the further action of light by being "fixed."

Fixing the images was a key element in the success of Talbot's concept. Earlier experimenters had made images using much the same chemical methods as Talbot. But all had failed to find a suitable fixative, and their images disappeared once they were exposed to daylight. Talbot's initial answer was to fix his prints in a strong solution of common salt, which rendered them insensitive to light.

Equally important was Talbot's grasp of the concept of a negative being used to make a positive. He noted in February 1835: "If the paper is transparent, the first drawing may serve as an object, to produce a second drawing, in which the lights and shadows would be reversed." The original appearance of the image would be restored, and multiple copies produced—a description of analog photography that serves as well today as it did then.

Events were unkind to Talbot. In France, independently, Louis-Jacques-Mandé Daguerre (1787–1851) had also been struggling to make images

An enlarged positive print from Talbot's Latticed Window *negative*

through the agencies of light and chemistry. The news of his discoveries was first reported in the *Literary Gazette* on January 12, 1839.

Talbot was put in a quandary. He had no details about Daguerre and his working methods or about Daguerre's process (*see* **Early Daguerreotypes of Italy**), which was actually different from Talbot's in that in Daguerre's process, images were made on metal—and not on paper. Still, as Talbot did not have this kind of detail at the time, he feared that Daguerre's invention would be identical to his own and would preempt any claim to priority that he could reasonably make. Time was against him; speed was essential. He selected examples of his work from the summer of 1835 to demonstrate "the wide range of its applicability." He mounted the photographs on slips of thin black paper to enhance their appearance. They were exhibited at the evening lecture of the Royal Institution on January 25, 1838, just thirteen days after the announcement of Daguerre's invention, and they were warmly received by the audience. Several years later, in 1844, Talbot had perfected his process to the point where he was able to efficiently develop lasting photographic images and place them on paper, the feat of which he displayed for the world in his work, *The Pencil of Nature*, which was the first book published containing photographic images.

In addition to *Latticed Window*, there are at least five surviving examples of other images similarly mounted, suggesting that they also were on display at the Royal Institution that evening in 1838. The only image to be dated and inscribed by Talbot is, however, *Latticed Window*. We do not know when or why Talbot added this annotation. The presence of this inscription has given the negative an exclusive identity, although it is possible that one of the other images from the 1835 collection predates it. Constantly reproduced in histories of photography, *Latticed Window* has an iconic status that will survive all subsequent revisions of the history of photography.

—ROGER TAYLOR

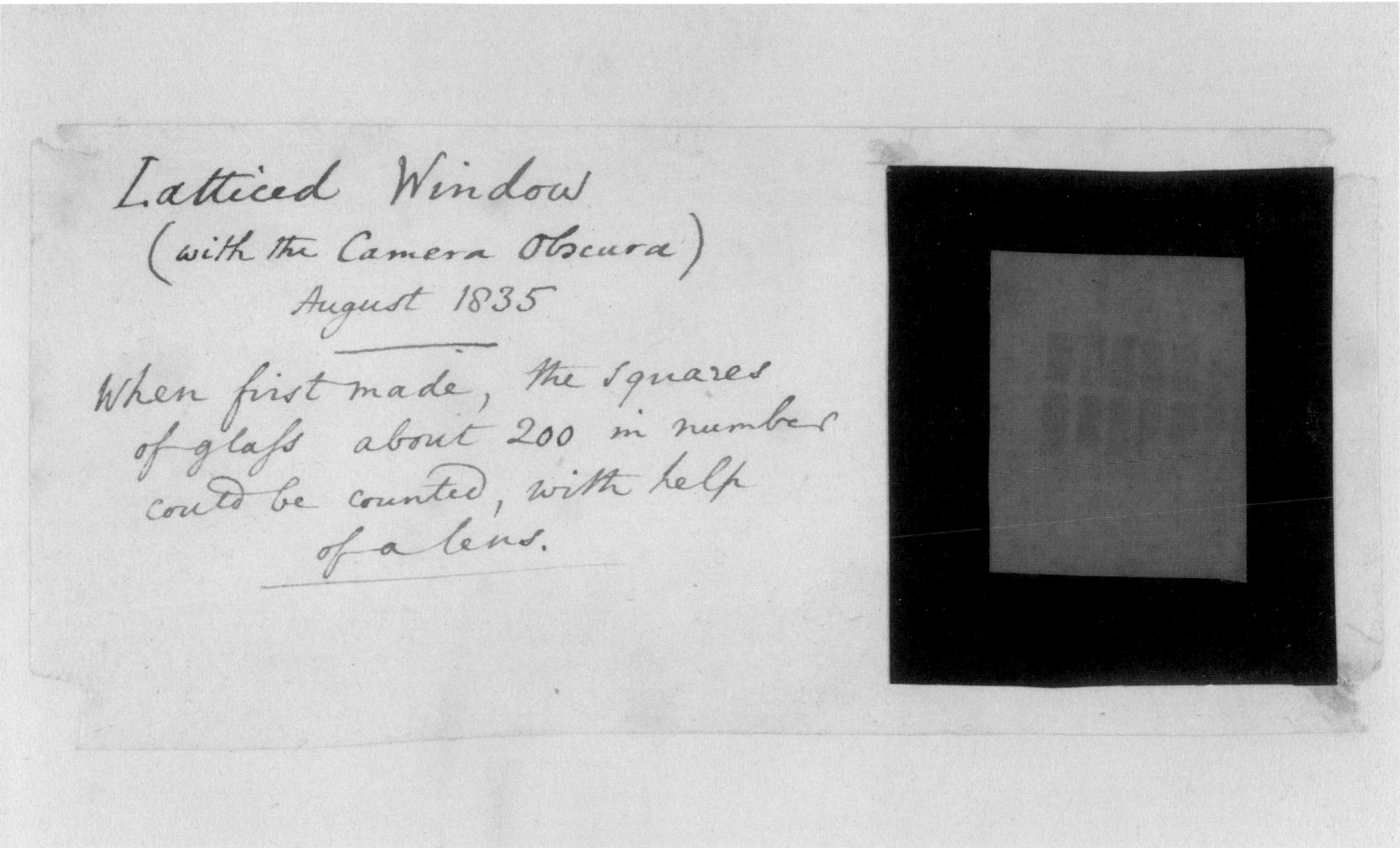

Latticed Window
(with the Camera Obscura)
August 1835

When first made, the squares
of glass about 200 in number
could be counted, with help
of a lens.

COOKE AND WHEATSTONE'S TELEGRAPH (1837)

In 1776, when the American colonies declared themselves independent of Great Britain, news of the event took forty-eight days to cross the Atlantic. Some sixty years later, an American named Samuel Morse found a way of eliminating such inconvenient delays. Variants of the Morse telegraph were later to link the world. The world's first commercially successful electric telegraph service did not, however, use Morse's invention, but was rather a development of Cooke and Wheatstone's telegraph, an instrument patented in 1837 by William Cooke (1806–79) and Charles Wheatstone (1802–75).

Inventors had been trying since the mid-18th century to use electricity for remote communication. All of them used what proved to be impractical means for sending and detecting signals. In this context, Cooke and Wheatstone triumphed. They used Hans Ørsted's then recent discovery that current electricity could move a compass needle. They added the idea of a coding scheme to make five wires do the work of twenty. And Cooke, in particular, realized that the railway network developing in Great Britain offered the ideal market for instruments of instant communication.

The instrument pictured opposite was made to demonstrate the principles that were laid out in the 1837 Cooke-Wheatstone patent. Five needles were used to transmit letters of the alphabet. A letter was transmitted when the operator pressed two of the buttons on the transmitting instrument, causing two of the needles on the dials of the receiving instrument to deflect in opposite directions and point to one of twenty letters. These movements also happened on the transmitting dial, as a guide for the operator. The use of twenty letters rather than the entire alphabet was a compromise between intelligibility and economy. Even with only twenty letters, five wires were required to connect the two instruments, one for each needle.

The idea of using a set of compass needles, and that their states of deflection would form a code, originated with Cooke, a retired army officer with no electrical training, in March 1836. His first experimental telegraph used three needles and six wires. Concerned about the number of wires needed, he then designed a "mechanical telegraph," a rather complicated system using a direct-reading clockwork receiver that needed only two wires to complete the circuit.

Cooke, however, had trouble getting his mechanical telegraph to work over more than a one-mile circuit. In February 1837, he met Wheatstone, Professor of Experimental Philosophy at King's College, London, who had experience with long electrical circuits. Wheatstone had also been experimenting with an electric telegraph, and he had designed a multi-button "permutating keyboard." Combined with Cooke's compass needles, the keyboard provided for the makings of a practical system. Wheatstone suspended the needles vertically in front of a lettered dial, so that the messages could be read easily—an important feature when demonstrating the telegraph to those who were unfamiliar with it.

Cooke and Wheatstone became partners in May 1837; their first patent was granted on June 12th. After an unsuccessful attempt to persuade Robert Stephenson, engineer of the London and Birmingham Railway, to invest in the construction of a telegraph along his entire line, they had success persuading Isambard Kingdom Brunel (1806–59), creator of the Great Western Railway, who had already seen a demonstration of their work. Cooke installed a system between Paddington, London, and West Drayton, about thirteen miles away; it was completed in July 1839. He used redesigned instruments with a rectangular dial and four needles, some letters being indicated by the movement of two needles (as before) and others by a single needle. Again, the trials were successful, but the equipment soon fell into disuse. This telegraph link was entirely reconstructed in 1843 and extended to Slough, using Cooke's new double-needle instruments. This was the first electric telegraph on which members of the public could send or receive messages for a fee.

In the meantime, in July 1840, Cooke had installed the world's first commercially successful electric telegraph on the London and Blackwall Railway. Single-needle instruments at each station communicated with multi-needle instruments at the engine houses and regulated the times the engines needed to operate. Enterprising telegraph operators developed their own private code for sending verbal messages on the single-needle instruments. When Cooke discovered what they were doing, he appropriated the idea for his 1845 patent for the single-needle system. The sequence of left and right deflections with a single needle made the convergence with Morse's dot-dash system almost complete. By 1870, the "speaking telegraph" had dropped Cooke's code in favor of Morse's—in which a dot was indicated by a left deflection and a dash by a right deflection. In this form it survived, and it stayed in use on Britain's railways until 1976.

The five-needle system operated with beguiling simplicity, but it was expensive to install because of the number of wires required. Contrary to nearly all historical accounts, its use was confined to only a few weeks of demonstrations that occurred between Euston and Camden in the autumn of 1837.

Acrimony developed between Cooke and Wheatstone over their respective shares of the invention. They were dissimilar characters who came together only when each had something to offer to the other. An arbitration settlement in 1841 quieted their wrangling for a while, but the partnership was dissolved in 1845. What matters more than these conflicts and rivalries is that Cooke and Wheatstone were the first to develop the means to speed public messages beyond the pace of ship or horse. It is for this accomplishment that they have been regarded as two of the pioneers of telegraphy.

—JOHN LIFFEN

THE BROUGHAM (1838)

Despite the abundance of carriage styles in the 19TH century, there were only a small number of fundamental changes in their general configuration. The most important of these changes were those that were introduced by the Brougham in 1838.

The design of carriages started to diverge from that of wagons with the introduction of suspension in the mid-16th century. In these early coaches, there was an undercarriage in which a large wooden beam called the "perch" joined the rear axle to the forecarriage, under which turned the front axle. At the four corners of the undercarriage, the body was suspended from upright posts by leather straps called "braces." This suspension improved as metal springs began to replace the braces in the latter part of the 18th century. It improved again with the introduction of the elliptical metal spring in 1804. However, the coach body was still high-mounted above the perch.

With the Brougham, this method was dispensed with, the body itself being all that linked the forecarriage to the rear axle. This change meant that the body not only became a structural element itself, but also that it now could be mounted down between the axles. Instead of a high, heavy, ponderous coach, this change made it possible to build a low, light, maneuverable, closed vehicle that could be drawn by a single horse. Such a carriage was better suited for use in cities and towns, and it was cheaper to buy and to operate than its predecessors. The Brougham, and the Clarence which developed from it, were the most popular kinds of closed carriage built, and they influenced the design of other closed- and open-carriage styles for the remainder of the century.

It is not known what gave Henry Brougham (1778–1868), Lord Chancellor of England, the idea for this new kind of carriage. It is known that he collaborated with a Mr. Robinson of the firm of Robinson and Cook of Mount Street, London, and that the carriage was delivered to him by the firm on May 15, 1838. In fact, it was not owned by Lord Brougham, but was on "job-bed"—contract-hired to him. Indeed, he is said not to have liked it very much and to have used it only occasionally. He later had an improved version made for him by the same firm. The original Brougham was sold to Sir William Foulis of Ingleby Manor, Stokesley, Yorkshire, on August 27, 1840. Compared with later models, it is rather boxy and inelegant; yet, it can be regarded as the prototype built before the proportions could be refined.

This Brougham subsequently became the property of Lord Henry Bentinck, who in turn sold it to Henry Earl Bathurst. It is reputed to have carried both Disraeli and Gladstone. Recognizing its historical importance, Earl Bathurst presented it to the Worshipful Company of Coachmakers and Coach Harness Makers in 1894, from which organization it has been on loan to the Science Museum since 1895. Mr. George N. Hooper of the Hooper Coachbuilding Company, a Past-Master of the Coachmakers Company, inspected the Brougham with Earl Bathurst at Cirencester House on April 5, 1894. He was able to confirm its origin by inspecting the books of the original builders and by speaking to Michael Elton, Earl Bathurst's former coachman, who had known the carriage in the Bathurst establishment for more than forty years. Hooper recorded his impressions of the Brougham in 1894: "It differs from those now made in many points of design—proportion, construction, finish. Those now made are much more refined in outline, and more pleasing to the eye; the body is several inches wider in front than at back, and although larger and heavier, does not provide the same comfort and convenience that those of smaller and lighter construction now afford."

—PETER MANN

Above: A c. 1903 print of a Brougham design. Opposite: The first Brougham was restored in 1977 for the tercentenary of the Worshipful Company of Coachmakers and Coach Harness Makers

During his visit to Venice, Dr. Alexander John Ellis (1814–90) rose early. Even as a novice photographer, Ellis already knew that early morning light offered the best results for his work. Before eight o'clock, he was busy setting up his camera, a cumbersome device that had to be placed on a tripod. After each exposure, he carefully recorded exposure times and weather conditions in his notebook—for future reference. There was still much to learn.

By mid-afternoon, when he reached the Rialto Bridge, Ellis had already made four exposures that he felt were "very good." The bridge was a natural subject for photography. Having gained access to an upper window of the White Lion Inn, Ellis carefully composed the picture and made his camera ready for the exposure. Exposure times were measured in minutes rather than in fractions of a second, and successful results were regarded as substantial achievements. Assessing the light, he judged that thirteen minutes would give a good result. These events took place on July 20, 1841, when the art of photography was in its infancy. Little more than two years had elapsed since Louis-Jacques-Mandé Daguerre and William Henry Fox Talbot had announced their exciting discoveries (*see* **Talbot's Latticed Window**).

Ellis was a fortunate young man. Born Alexander Sharpe, he had changed his name in 1825 to satisfy the terms of a bequest from a relative who wished him to devote his life to study and research. Educated at Shrewsbury and Eton, he became a scholar at Trinity College, Cambridge, in 1835. After graduating in 1837, he entered the Middle Temple, but he appeared to have no intention of devoting his life to law. Travel was infinitely more appealing to a young man of wealth and leisure.

In March 1839, Ellis was in Dresden, where news of an exciting discovery by Monsieur Daguerre of Paris was appearing in the newspapers. Skeptics could not believe that it was possible to make images by the agency of the sun alone.

Daguerre's photographic process was accomplished using solid metal. A highly polished sheet of silver-plated copper was made sensitive to light by exposing it to the fumes of iodine. After exposure, the image was "brought out" by further exposure to the fumes of warmed mercury. The resultant plate, known as a daguerreotype, drew gasps of amazement from an admiring audience when it was displayed. What captured the imagination were the exquisite detail and the infinite accuracy of the visual information, which was beyond anything that the hand of man could produce.

The traditions of the 18th-century Grand Tour still lingered in early Victorian Britain, despite the complete abandonment of such touring during the Napoleonic Wars, when the whole of Europe was effectively closed to leisured travelers. Even so, interest in the sites of the Ancient World, seen as the cradle of European civilization, was still strong among the educated classes. It was a necessary part of their upbringing to understand the history and culture of these early civilizations and to follow the examples of ancient Greece and Rome.

In London, print-sellers offered engravings of the classical sites, though they were frequently based more on imagination than on truthful observation. If accuracy mattered, there was no better medium on which to base the engravings than the daguerreotype. The French had already led the way with *Excursions Daguerriennes*, in which highly detailed engravings were based on daguerreotypes by Noël-Marie-Paymal Lerebours (1807–73).

Using this model, Ellis planned to publish a series of sixty engravings entitled *Italy Daguerreotyped*, which he believed would offer "that verisimilitude which is required by the traveler who wishes for reminiscences, the antiquarian who desires historical evidence of the state of buildings as existing at known times, and to all who wish to know what the buildings of Italy really are."

He arrived in Pozzuoli on May 21, 1841, and began a photographic odyssey through Naples, Pompeii, Rome, Assisi, Pisa and Florence, arriving in Venice on July 14th. During those eight weeks he recorded 137 exposures in his notebook. It would be easy to underestimate this achievement in comparison with today's photography, where all labor and most skill are sublimated solely to technology. For Ellis, the burden of the camera, lenses, metal plates and chemistry would have been considerable. As his intention was to publish accurate engraved transcriptions of the plates, size was an important factor. Smaller, more convenient plates would not provide the engraver with adequate information and detail. The practice was to have the engraving plate the same size as the daguerreotype. Despite the added inconvenience, Ellis chose large 150 x 205 millimeter plates to ensure high-quality results at every stage of his project.

There was a small amount of published material on other photographers' experience, which he could have drawn on for guidance. Still, that guidance would have been limited. All exposures had to be estimated by trial and error. The climate affected results. In the heat of an Italian summer, the chemicals acted variably, and Ellis' patience and skill were severely tested. A lack of understanding also led to frustration and failure. In Venice, on July 17th, he repeatedly tried to make a successful exposure. After eight attempts, he attributed his lack of mastery to the damp carpet in the bottom of the gondola on which the plates had been placed for safekeeping. In Rome, he collaborated with two Italian photographers, Achille Morelli and Lorenzo Suscipi. From them, he bought multi-plate panoramic sequences of the city from the Capitol Tower and San Pietro in Montorio. He acquired from them an additional forty-five plates to add to those he had made himself.

However, his scheme to publish—planned for January 1, 1845—came to nothing. It was, perhaps, a casualty of changing tastes at a time when novelty was paramount.

The surviving 159 daguerreotypes are the largest and most comprehensive group of topographic daguerreotypes known—and they are almost certainly the earliest photographs of Italy.

—ROGER TAYLOR

Opposite: Venice, Bridge of the Rialto and Rive de Carboni from the White Lion Inn, *by Alexander John Ellis, 1841*

THE ROSSE MIRROR (1842)

Even in the 21st century, when we are accustomed to scientific apparatus on a large scale, the Earl of Rosse's six-foot mirror is striking for its size. It is also striking for its workmanship, and when its history and context are understood, it becomes a very significant object.

William Parsons (1800–67), the 3rd Earl of Rosse, had as his family's seat Birr Castle in what is now County Offaly, Ireland. From the 1820's he devoted considerable time, energy and wealth to the study of astronomy. He was particularly interested in the construction of large reflecting telescopes.

In the late 18th century, Sir William Herschel had pioneered the production of large reflecting telescopes and made many remarkable discoveries (*see* **Herschel's Seven-Foot Telescope**). Herschel had abandoned refractors—telescopes using lenses—for reflectors, because good lenses could not be made large enough to provide the required light-gathering power. By the 1840's, however, the situation had changed. There had been major developments in the production of flint glass and in the design of lenses. The new observatories of the early 19th century were equipped with refractors that were easy to maintain and ideal for the positional astronomy currently in favor. Despite this circumstance, the Earl of Rosse was determined to pursue his line of interest and continue in William Herschel's tradition.

He began by erecting a foundry and workshop on the grounds of Birr Castle, overcoming the shortage of skilled labor by training tenants on his estate. His first mirrors were six inches in diameter, followed by fifteen, twenty-four and, eventually, in 1840, thirty-six inches. The three-foot mirror was a great success, encouraging Rosse to go beyond Herschel's limit of four feet and to plan a six-foot reflector.

The casting, grinding and mounting of a mirror of this size was a tremendous challenge to the technological capabilities of the time. The foundry had to be enlarged to accommodate the three furnaces needed to melt sufficient metal to fill the mold. An alloy known as speculum metal, consisting of approximately two parts copper to one of tin, was used. The mold was made by binding together layers of hoop-iron arranged to allow air to pass through so that the metal cooled more quickly.

On the night of April 18, 1842, a crowd gathered to watch the first casting. The crucibles, holding one ton of metal each, were subject to nineteen hours of intense heat created by using turf fires. As the molten metal was poured, the crowd saw "a burning mass of fluid matter…settling into a monument of man's industry for ever." The mirror, still molten, was dragged along a railway track to the annealing oven, where it was cooled gradually for sixteen weeks. The grinding and polishing could then begin. Five mirrors were made in this way, only two of which were suitable for use. The Rosse mirror pictured opposite was the first successful one.

The mounting of the telescope proceeded throughout 1843 and 1844. The mirror had a focal length of fifty-four feet, and the enormous telescope tube was supported between two walls fifty feet high, seventy feet long, and twenty-three feet apart. The tube was so large that one of Lord Rosse's friends, Dean Peacock, was able to walk through it

The "Leviathan of Parsonstown"

under an open umbrella. The walls ran in a north-south direction along the meridian, and, because the telescope had very little lateral motion, it was effectively a transit telescope, that is, one that views celestial objects as they pass over the meridian.

The telescope became known as the "Leviathan of Parsonstown," and it had the distinction of being the largest in the world for seventy-five years. From 1848, the telescope was trained almost exclusively on nebulae, luminous clouds that had been seen by earlier astronomers. In 1850, the Earl of Rosse reported that, through his telescope, the great nebula in the constellation Andromeda could be seen to have a spiral structure. We now know that it is a spiral galaxy like our own. He was able to see stars in so many nebulae that he became convinced that all nebulae were in fact star clusters, a theory that was later disproved.

Although the telescope was well used for thirty years, there were difficulties, the principal one being the Irish weather. Lord Rosse and his colleagues spent many frustrating nights waiting for clouds to disperse, and the constant dampness tarnished the mirror. Another problem was that four people were needed to operate the instrument, which took about ten minutes to set up and was not easy to move. A third difficulty was that the Earl of Rosse had other commitments: he was Member of Parliament for the area during the difficult famine years, and he also served as President of the Royal Society in the years 1848–54.

The endeavors of the Earl of Rosse demonstrated to the scientific community that the making of large reflectors was not impossible. When the technique of coating glass was discovered in the second half of the century, reflectors gradually returned to the forefront of astronomical discovery. Many of Rosse's techniques were then copied and developed further.

—JANE WESS

ELIAS HOWE'S SEWING MACHINE (1846)

"One of the few useful things ever invented" was Mahatma Gandhi's assessment of the sewing machine. Up to 1845, there had been several aspirants to the title of "inventor" of such a device, yet each machine that was created was flawed in some way. In that year, however, Elias Howe, Jr. (1819–67) triumphed in incorporating the most significant ideas to date into a machine that became a turning point for the sewing machine and thereafter for the mass-produced clothing industry. Born in Spencer, Massachusetts, Howe was the son of an English immigrant farmer. He succeeded where others had failed, not only because of his ingenuity, but also because his work was timely. His work fit neatly into a period when machine-made cloth had become commonly available, and engineering had advanced sufficiently to allow for the manufacture of affordable machines of great complexity.

Rather than face the mechanical nightmare that imitating manual sewing would have incurred, Howe bypassed the problem by using the "eye-pointed" needle and the lockstitch. This needle was devised by Londoner C.F. Weisenthal, in 1755, for embroidery. The lockstitch combined two separate threads, one on each side of the fabric; it was developed in the 1830's by Walter Hunt of New York. Luckily for Howe, Hunt did not patent the lockstitch, so anyone could use his idea.

In the Howe sewing machine, an eye-pointed needle penetrates the cloth only just far enough to pass the thread through. On retracting, the friction of the cloth on the cotton thread causes a small loop to remain on the other side of the cloth, just large enough for the second thread to be passed through. Howe used a curved needle, swinging sideways in an arc. The secondary thread was carried in a shuttle; in this way, Howe had applied one of the principles of weaving to seam sewing. Unlike modern machines, the cloth in Howe's machine hung vertically from pins along the edge of a plate; it edged forward, stitch by stitch. After a few inches' sewing, the plate had to be reset and the garment re-pinned in a new position.

The Howe sewing machine, showing cloth pinned up for sewing

Having been granted a United States patent in 1846, Howe looked to England for a wider market for his invention. A machine was taken to England, where William Thomas of Cheapside, an umbrella and corset maker, became interested in its promise of greater productivity. An agreement was concluded whereby Thomas patented the machine in England and in return paid Howe royalties. But the English venture turned out to be a disaster. The arrangement with Thomas did not work out (for Howe, at least), and Howe was beset by debt and poverty. His wife became ill with tuberculosis, from which she died in 1849. Back in America the same year, Howe found that the technology and application of sewing machines had advanced; he also felt that other sewing-machine makers had been using his patented ideas.

Howe embarked on a series of legal battles, challenging makers whose machines incorporated features that he claimed as his own; he even took on the formidable Isaac Singer (1811–75) and won. Patent battles tied up the entire sewing-machine industry, and progress ground to a halt. But by 1856, Howe had been persuaded that peace rather than conflict was the way ahead, and he accepted an agreement whereby all significant patents would be pooled for the benefit of all concerned. Howe's fortune was, at last, assured; it is estimated that, by 1867, when his patent expired, he had earned $2 million. Sadly, he did not live to enjoy his fortune.

The machine pictured opposite has direct links with Howe's ill-fated London venture. It was given to the Science Museum in 1919 by the son of William Thomas, who stated that it was the machine that Howe brought from America to England in November 1846.

The growth of the sewing-machine business, of which Howe was one of the founders, turned out to have profound industrial and social implications. The labor-saving possibilities alone were enormous; in 1861, a shirt took just over an hour to make by machine, compared to fourteen hours by hand. Mechanized sewing spread to many products other than clothing: ships' sails, cloth bags and sacks, and boots and shoes. The power-driven needle was as welcome as mechanization itself.

The reception of sewing machines into the home occurred in the United States before it did in England. Singer, in particular, saw the potential of selling domestic machines in huge quantities: his marketing was aggressive, and he pioneered time-payment and rent-to-own schemes in order to penetrate the lower end of the market. As with so many labor-saving devices, acceptance in England occurred some years later, due to the fact that people wealthy enough to buy machines could employ servants to do this kind of work for them—using old-fashioned methods. In contrast, the United States enjoyed fuller employment and, perhaps, more democratic attitudes.

—ERYL DAVIES

JOULE'S PADDLE-WHEEL APPARATUS (1849)

James Prescott Joule (1818–89) was born in Salford, a part of the Manchester metropolitan area that was then the center of cotton production in Great Britain. The cotton industry relied on steam power, and Joule's work laid one of the foundations of the science of thermodynamics, which provided a scientific basis for the development of the steam engine and other heat engines.

Joule's father was a wealthy brewer, and for much of his life, Joule was able to indulge his interest in science in a private laboratory. His earliest research was in electromagnetic engines, the primitive electric motors that seemed to promise an inexhaustible supply of motive power. Joule advertently proved to himself that an electromagnetic engine was much more expensive to run than a steam engine, so he began instead to investigate the heat produced by electric current. He quickly formulated a law relating the quantity of heat produced to the current and to the resistance of the conductor in which it flows. This was a most significant discovery, but Joule went further: he began to challenge the then accepted theory of heat.

There were two rival theories about the nature of heat: the "caloric" theory—that it was a subtle fluid; and the "dynamical" theory—that it represented the motion of the atoms of matter. Either theory, with some difficulty, could explain known thermal phenomena. In both theories, it was assumed that the total heat of a system of bodies was conserved. Thus, for example, when two objects were rubbed together, they became hot because heat was transferred from somewhere else in the system, not because the friction actually generated the heat.

Joule performed a series of electrical experiments in which he proved, first of all, that heat was produced by an electric current, not merely transferred from another part of the electrical system. Then, he showed indirectly that heat could be converted into mechanical work. Finally, he measured the mechanical work involved in producing a unit of heat.

Joule reported his results at a meeting of the British Association for the Advancement of Science in 1843, but his findings aroused no interest. The idea that heat was conserved was too strongly established to be shaken by a little-known person working outside established scientific institutions, and many of his listeners questioned the reliability of Joule's conclusions, drawn as they were from a few measurements of small temperature changes.

Undaunted, Joule built new apparatus in which mechanical work was converted into heat directly, by stirring water vigorously with a horizontally rotating paddle wheel. The work needed to rotate the paddle wheel was provided (as it was in his previous experiments) by weights falling through a measured distance. Joule reported his new measurements to a British Association meeting in 1845, again with no obvious impact, then presented further measurements to a meeting in 1847. This time, his ideas created much more interest, largely because they attracted the attention of the young William Thomson (1824–1907), who, as Lord Kelvin, was to become one of the great physicists of the age.

Thomson did not immediately accept Joule's work, but in the end it led him, along with men such as Clausius and Rankine, to accept the theory of thermodynamics. Joule's acceptance was complete when the Royal Society, in 1850, published his account of new and more accurate experiments.

Joule did not have the mathematical skill to contribute further to thermodynamics; indeed, some of his good fortune was due to his meeting William Thomson. In Germany, J.R. Mayer had proposed similar ideas in 1842. He too had measured the mechanical value of a unit of heat, though less rigorously than had Joule. Mayer's work was mostly ignored.

Joule's research established the first law of thermodynamics, the principle of the conservation of energy, though it was the German physicist Hermann von Helmholtz (1821–94) who, in 1847, formulated this general conclusion: energy is conserved, not heat; heat is one form of energy, and it may be converted into another. Joule's work extended the dynamical theory of heat, which he had always supported, and demolished the caloric theory, which became untenable once it was known that heat is not conserved.

The "mechanical equivalent of heat" remained important for many years, and it was featured strongly in school physics. Not until the 1960's and the introduction of the International System of Units was Joule's work taken to its logical conclusion. The mechanical equivalent of heat is now a historical curiosity because all kinds of energy are measured by the same unit. That unit, first defined in 1889, is called the "joule."

—Neil Brown

ORIGINAL MAUVE DYE (1856)

"If your discovery does not make the goods too expensive, it is decidedly one of the most valuable that has come out for a very long time. This colour is one which has been very much wanted in all classes of goods and could not be obtained fast on Silks, and only at great expense on cotton yarns."

So wrote Robert Pullar of John Pullar & Sons, a leading firm of Scottish dyers, on June 12, 1856, following evaluation of fabric samples treated with mauve dye developed by the teen-aged William Henry Perkin. Some eighteen months later, this novel chemical product was being dispatched from the Greenford Green factory of Perkin and Sons to the largest silk dye works in London. Not only was this the first step in the industrialization of organic chemistry, but it was also the beginning of the commercialization of scientific invention.

Before Perkin's time, virtually all dyestuffs were of vegetable or animal origin, and the choices had hardly increased since the Middle Ages despite efforts at improving methods of dyeing as a result of the vast growth of textile manufacture during the Industrial Revolution. The context of Perkin's discovery was, however, quite unrelated to these efforts. W.H. Perkin (1838–1907), the son of a builder, showed a keen interest in chemistry from an early age. In 1853, he enrolled at the Royal College of Chemistry, where he attended the lectures of the German chemist August Wilhelm von Hofmann (1818–92), renowned for both his research and his teaching abilities.

In the middle of the 19th century, quinine was much in demand for the fight against malaria, but it was expensive. Hofmann and other chemists investigated alternatives. From his very limited knowledge of chemical constitutions, Hofmann speculated in 1849 that quinine might be synthesized from naphthylamine, a substance obtainable from coal tar.

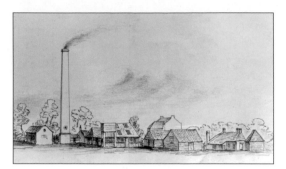

Perkin's factory at Greenford Green near London

In 1856, Hofmann's young research assistant, Perkin, resolved to attempt a synthesis from the oxidation of allyltoluidine, which was also to be obtained from coal tar. Perkin undertook this oxidation at home, having in his enthusiasm for research equipped part of a room in his father's house for this purpose soon after he had begun his studies with Hofmann. Working during the Easter holiday of 1856, he did not succeed: his oxidation yielded none of the desired colorless quinine but only a dirty reddish-brown sludge. He repeated the oxidation, this time with the simplest base available from coal tar, aniline. The result was a black precipitate, which, upon drying, and after being treated with methylated spirit, yielded an intense purple solution. To his surprise, Perkin found that it dyed silk a beautiful color; also, it better resisted the fading effects of light than did existing dyes.

The eighteen-year-old Perkin pursued his discovery with the vigor of youth and with the wise counsel of those in the dyeing industry with whom he was put in contact. He deposited a provisional patent in London on August 26, 1856, and the final version was sealed the following February 20th. His resignation from his post at the Royal College of Chemistry in October—which Hofmann regarded

as foolhardy—further testified to Perkin's commitment to make money from his synthetic dye. Producing the dye on a commercial scale required Perkin's father to recognize that his son's interest in chemistry had not been in vain, as he once had feared, and that the necessary capital expenditure to back his son might be a sound family investment. This decision to do so proved to be a wise one.

Perkin's discovery gave impetus to a new coal-tar/dyestuffs industry in which the level of protection afforded by patents was much less than it is today. Fortunately for Perkin, although slight modifications of the original process were introduced, none outdid the economy of the original method. As well, Perkin kept up his own research activities, and he introduced new coloring materials, notably Britannia violet (derived from magenta) in 1864. His ingenuity helped to keep the Greenford factory operating at a profit as more brilliant dyes displaced mauve from the market after a span of less than ten years.

In 1869, Perkin devised two new methods that allowed for the industrial manufacture of alizarin (the natural coloring substance contained in madder, which was the primary red dye of the period), the synthetic of which had been reported by Graebe and Liebermann in 1868. By the end of 1869, Perkin's company had made its first ton of alizarin, expanding output to more than 200 tons per annum by 1871.

Perkin had, however, always hoped to devote himself to pure research, and in 1873, at the age of thirty-five, he sold his interests and "retired." His early work was the foundation of the artificial dyestuffs industry we know today. However, British industrialists failed to build on Perkin's pioneering work. As a result, Germany, whose scientists were in the forefront of new branches of chemistry, soon took the lead in the production of artificial dyes.

—Derek Robinson

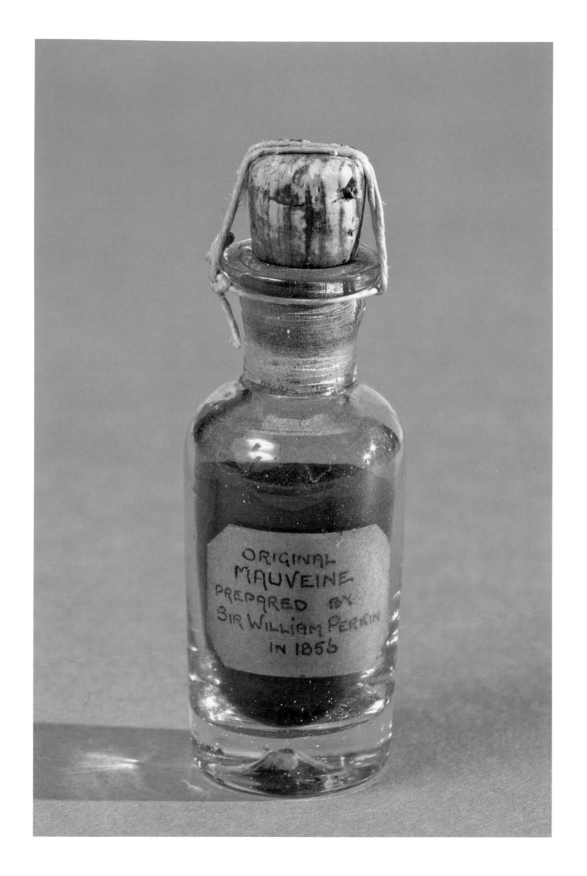

Early in 1855, when photography was not yet twenty years old, an Oxford student, Charles Lutwidge Dodgson (1832–98), son of the rector of the Yorkshire village of Croft, went to the nearby town of Ripon and "…got my likeness photographed by Booth. After three failures, he produced a tolerably good likeness, which half the family pronounce the best possible, and half the worst possible."

Dodgson is today best known by the pen name "Lewis Carroll," under which he wrote two of the most enduring of children's books, *Alice's Adventures in Wonderland* and *Alice Through the Looking-Glass*. His Ripon portrait was taken with the wet collodion process, which its inventor, Frederick Scott Archer, did not patent, but made freely available to all. Unlike the earlier calotype and daguerreotype, it had made photography relatively cheap, reliable, and accessible.

One of Carroll's uncles, Skeffington Lutwidge, was an amateur practitioner. Later in 1855, when Carroll was on summer vacation from Christ Church, Oxford, he observed his uncle's variable results and wrote *Photography Extraordinary*, the first of his many works to use the medium as subject matter. It playfully proposed an imaginary camera that would turn a writer's half-formed ideas into a finished novel.

Back in Oxford, after taking advice from Lutwidge, he and a fellow student/enthusiast, Reginald Southey, traveled to London in March 1856 to buy a camera at Ottewill's, in Islington.

Though the wet collodion process was easier to manipulate than the daguerreotype or calotype, the photographer, to make a negative, had to mix, precisely and skillfully, his own chemicals (in the dark) and spread them evenly over a sheet of glass. If he did this carelessly, or left any part of the glass uncovered, he got a faulty picture as a result. He had to place the negative in the camera, and take the picture, while the chemicals were still moist (hence "wet" collodion); when dry, they lost their photosensitivity. Not surprisingly, there were many failures.

After photographing buildings, statues, medical specimens, trees, gardens, and still-lifes, Carroll

Lewis Carroll: self-portrait probably taken at his rooms, Christ Church, Oxford. Opposite: Reginald Southey and Skeletons, *1857*

turned to portraiture. He began a remarkable series of photographs of university colleagues and friends, who posed in his rooms in Christ Church or outside in the college gardens. In the photograph shown opposite, Reginald Southey was clearly posed outdoors, apparently in front of a sheet of plain wood, erected for the purpose. Southey was a medical student, and the skulls and skeletons with which he is standing suggest that he had views on Darwin's theory of evolution, the subject of furious debate at Oxford during that time.

Later, Carroll constructed a studio—or "glass house"—on the roof of the main ("Tom") quadrangle of his college, to which he brought hundreds of sitters. For twenty-five years, he seems to have spent more time on photography than on his writing or on his work as a mathematics tutor.

Increasingly, Carroll's subjects were young girls. Alongside his diary entries for March 1863,

he listed 103 girls whom he wanted to photograph, in alphabetical order of their first names. He must have known them all intimately, at least in his imagination. It is a real achievement to have persuaded those who did pose for him to keep so still in front of his camera. The shortest exposure time he recorded is forty seconds, and his diary is as full of notes of failed photographs as Victorian family albums are full of blurred faces, hands and eyes. But Carroll's successes made him, as the distinguished photo-historian Helmut Gernsheim has said, "the greatest Victorian photographer of children."

Though Carroll's pictures of adults and children are less stilted and artificial than many in the Victorian era, they are nonetheless carefully considered. Poses and groupings were his "favourite branch of the subject," and he concentrated particularly on "the arrangement of the hands." Because of his long exposure times, sitters are virtually never shown smiling, but this photograph—*Reginald Southey and Skeletons*—has something of the humor found in Carroll's books.

The mystery about Lewis Carroll's photography is that he abandoned it in July 1880, without warning or explanation. There are several possible reasons. One is that he disliked the dry plates that were replacing wet collodion. Another is that his enthusiasm for photographing young girls, preferably in very few clothes, finally got him into trouble. But the most convincing reason is somewhat less sensational.

As the sale of Carroll's books grew, he may well have decided to concentrate on writing. A year later, he resigned his mathematics lectureship: "I shall now have my whole time at my disposal and, if God gives me life and continued health and strength, may hope, before my powers fail, to do some worthy work in writing." As he approached fifty, he turned to his real vocation, and he is now regarded as one of the great humorists of English literature.

—COLIN FORD

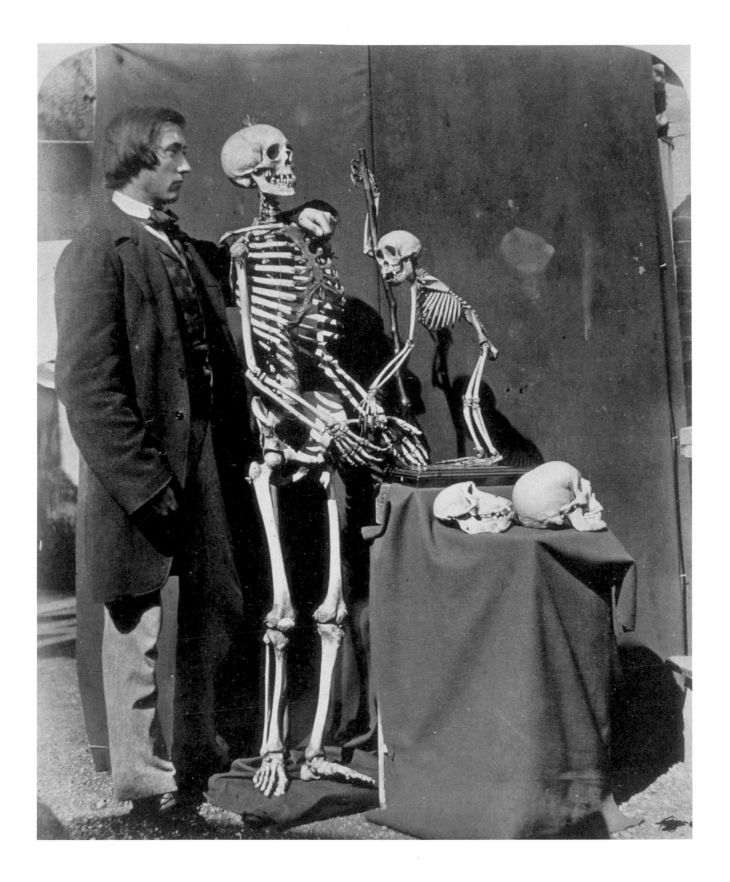

THOMSON'S MIRROR GALVANOMETER (1858)

Thomson's mirror galvanometer was the only instrument that could reliably detect signals sent through the first transatlantic telegraph cable, laid in 1858. The instrument shown opposite was used at the Newfoundland end of the cable.

The galvanometer, an instrument for detecting and measuring electric currents, was developed by the Italian scientist Leopoldo Nobili (1784–1835) in the 1820's. It was the first practical application of the discovery, announced by Øersted in 1820, that an electric current in a wire produces a magnetic field. In a galvanometer, the current being measured is passed through a coil, creating a magnetic field. The magnetic field interacts with a magnet, which can turn (in some later instruments the magnet is fixed, and the coil turns). The extent to which the magnet turns is a measure of the current flowing in the coil. Usually the magnet carries a pointer that moves across a scale, and in some instruments, the magnet and the pointer are the same piece of iron.

The electric telegraph began to be exploited commercially in the 1840's. Nearly all telegraphs used receiving instruments that were essentially galvanometers. When an electric current was sent through the telegraph wire, the galvanometer needle at the other end moved in response. In most telegraphs, the messages were comprised of sequences of long and short current pulses, using Morse or a similar code.

Most inland telegraph wires were carried on poles in the open air, although underground cables were sometimes used for short distances in cities. To establish telegraph connections with other countries, Britain adopted submarine cables. By the 1870's, the time for messages to travel from continent to continent was reduced from days or months to just hours.

Long submarine cables have very different electrical characteristics from wires carried on poles and spaced apart in the air, and, as a practical consequence, signaling becomes very slow. In 1858, the electrician for the Atlantic Telegraph Company, Wildman Whitehouse, attempted to solve the

Sir William Thomson, Lord Kelvin

problem by using very high voltages from an induction coil to send the signals. The result was damage to the cable.

The problem was solved by William Thomson (1824–1907), the Professor of Natural Philosophy at Glasgow University, who was already well known for his work in electricity. He appreciated that the solution to the problem of sending a detectable signal through a cable was to use a small current and a very sensitive galvanometer as a current detector. With a small current, the charging time was reduced, but the problem of detecting the current was greatly increased.

Thomson's galvanometer operated in the same way as all other galvanometers, but its construction was very light. The magnet was small, and, instead of having a pointer, it carried a tiny mirror, fixed to the magnet. A beam of light shone on the mirror, and the slightest movement of the magnet and mirror would cause the reflected beam of light to move across a scale. The system could detect much

smaller currents than any previous meter, and it made possible the first transatlantic telegraph in 1858.

Thomson was knighted for his contributions to the development of the electric telegraph; he was subsequently raised to the peerage as Baron Kelvin of Largs, the first person to be so honored for contributions to science.

The 1858 cable lasted only a short time (there had been manufacturing and cable-laying problems, as well as the damage caused by Whitehouse's experiments). When a more successful cable was laid in 1866, the electrician Latimer Clark arranged a striking demonstration in which a simple electric cell consisting of a silver thimble, a small piece of zinc and acid, sent a current across the Atlantic and produced a strong deflection in a mirror galvanometer.

Two people were needed to receive Morse code signals by means of the mirror galvanometer: one watched the beam of light, noting the Morse dots and dashes, calling out the letters for the second to write down. By 1867, however, Thomson had developed an improved instrument for receiving signals, his "siphon recorder." It was still basically a galvanometer, but instead of carrying a mirror, the magnet carried a small, siphon-shaped glass tube that had one end in a vessel of ink and the other end close to a moving strip of paper. As signals were received, the end of the siphon tube moved sideways across the strip of paper, which was drawn through the receiver. The siphon recorder was also a very sensitive instrument, and because it made a permanent record on the paper of the messages received, it was no longer necessary for the instrument to be watched continuously.

Although mirror galvanometers were soon replaced as telegraph instruments by Thomson's siphon recorder, they continued to be used for more than a century, whenever electrical measurements at high sensitivity were required.

—Brian Bowers

THE KEW PHOTOHELIOGRAPH (1858)

The sun is not the perfect body that Aristotle would have had us believe: it is occasionally flawed by the presence of spots. It is curious that, although sunspots had been observed since the early days of the telescope, it was not until more than 200 years later that the German amateur astronomer Samuel Heinrich Schwabe (1789–1875) discovered that their occurrence is cyclic. They wax and wane over a period of about eleven years, although later research has identified longer cycles as well.

It was only natural that British astronomers would want to conduct regular solar observations to confirm and extend Schwabe's findings. Sir John Herschel (1792–1871) was particularly vocal in demanding a suitable instrument for this work, and in 1854 he was able to report to the Kew Committee of the British Association for the Advancement of Science that £150 had been offered by the Donation Fund of the Royal Society for that purpose.

The King's Observatory at Kew had been built so that George III could observe the Transit of Venus in 1769, but the British Association took on its maintenance in 1842, in part as "a repository and station for trials of new instruments." In the beginning, these instruments were primarily meteorological and magnetic.

As a pioneer of astronomical photography, using a technique that employed wet collodion plates, Warren De La Rue (1815–89) was a natural choice to supervise the construction and use of the new solar instrument. The Kew photoheliograph was commissioned from the optical-instrument maker Andrew Ross: a combined camera and telescope with a three-and-a-half-inch lens of fifty-inch focal length, a spring-loaded shutter to permit rapid photographs of the sun and a clockwork drive to compensate for the earth's rotation. The Kew photoheliograph was first used in 1858, in time to make routine observations around the minimum of the solar cycle. Only two years later, it was dismantled so that it could participate in the observation of an eclipse of historic significance. This participation involved moving the photoheliograph—to Spain.

It can have been no mean feat to transport the photoheliograph, plus its cast-iron mounting and sundry photographic apparatus, by sail and stage-coach to a village in northern Spain. But on two counts De La Rue was convinced that this journey was worthwhile. First, the solar eclipse of July 18, 1860 would be the first opportunity to test wet-collodion-plate photography at a solar eclipse (earlier photographic attempts had produced meager results). Second, De La Rue wanted to settle a scientific question by photographic means.

The question had to do with the nature of the controversial plumes seen on the limb of the sun in eclipse, when its blinding light is obscured momentarily by the moon. Did these "prominences" arise in the earth's atmosphere or in the sun's? A relatively simple way to settle the matter would be to record the eclipsed sun from different locations. If prominences were some new meteorological phenomenon, observations from widely separated sites ought to differ, but if prominences were solar in origin, they should match.

De La Rue established his observatory on a threshing floor in Rivabellosa, on the borders of Rioja country. Meanwhile, Father Angelo Secchi (1818–78) set up his telescope at Disierto de las Palmas, about 300 miles to the southeast, near the Spanish coast. De La Rue was able to take thirty-five usable photographs, including two crucial ones, during the precious minutes of total eclipse.

These photographic plates, wrote De La Rue, "depicted the luminous prominences with a precision as to contour and position impossible of attainment by eye observations." Although the faint solar corona visible during the eclipse eluded photographic record, a plethora of prominences appeared. When De La Rue subsequently compared Father Secchi's photographs with his own, he was thrilled to find that they coincided. There was no longer any question: prominences belong to the sun itself.

Back in England, the photoheliograph provided a series of photographs that covered an entire solar cycle. It took its last solar photographs at Kew at the end of March 1872.

—JOHN DARIUS

The temporary observatory at Rivabellosa. The lens of the photoheliograph is just visible above its shelter. Warren De La Rue appears twice; this view was montaged from two separate photographs

THE FIRST PLASTIC (1862)

Plastics are now very familiar to us. They can be used for an almost infinite variety of purposes, from fabrics, furniture and crockery, to cars, airplanes and space shuttles. What is generally accepted as the first plastic, Parkesine, was formulated about 150 years ago.

A plastic is a material that can be molded or shaped into different forms under pressure and/or heat. Chemically, plastics consist of long chains of molecules made up predominantly of carbon and hydrogen atoms. The great advantage of plastics over other materials is that they can be used to create almost any kind of manufactured goods.

The earliest plastics, beginning with Parkesine, were semi-synthetic, based on natural substances such as cellulose obtained from sources such as cotton flock or wood fibers. Parkesine dates back to the middle of the 19th century, a time when much work was being done to find materials that would make satisfactory replacements for materials such as ivory, amber and tortoiseshell, which were becoming increasingly rare and expensive.

Parkesine was the invention of an energetic Birmingham scientist, Alexander Parkes (1813–90). He extended the work of a Swiss chemist, Schönbein, who, in 1845, had produced pyroxylin (cellulose nitrate) from cellulose fibers mixed with nitric acid. Parkes discovered that if he mixed cellulose nitrate with plasticizers and solvents, he produced a moldable substance which he called Parkesine. He exhibited it in a variety of forms at the International Exhibition of 1862, including combs, hair ornaments/holders, billiard balls, and carved plaques. Parkes' product so impressed the judges at the exhibition that he was awarded a bronze medal for "excellence of product."

Parkes went on to produce a wide range of objects with Parkesine. They involved a variety of uses, as well as rich jewel-like colors, and included fishing-rod spools and a miniature, intricately carved head of Christ crowned with thorns, in imitation amber.

The success of Parkesine at the International Exhibition of 1862, and the fact that he believed

Alexander Parkes: lithograph by Abraham Wivell, Jr., 1848

that he could reduce the production price to one shilling (five pence) a pound, encouraged Parkes to establish the Parkesine Company Ltd. in 1866 with capital of £10,000. Although it had been created with great enthusiasm by Parkes, the company soon foundered, and in 1868 it went into liquidation. Parkes himself blamed the flammability of the material. Others have suggested that the Parkesine Company failed as a result of Parkes' obsession with keeping the price low, and hence, perhaps, using materials that were cheap and of poor quality. Maybe he would have succeeded had he marketed Parkesine as a luxury product at a higher price.

Daniel Spill (1832–77), who had been Parkes' factory manager, then attempted to make a commercial success of Parkesine, which he renamed Xylonite. Spill established the Xylonite Company in 1869 to produce and market this product. At almost the same time, John Wesley Hyatt was carrying out similar work in the United States. The earliest cellulosic plastics (both in the U.S. and Great Britain) had been plagued by the problem of flammability. Fires were a continuing problem with plastics based on cellulose nitrate, and the problems were never totally resolved.

Hyatt realized that camphor made an excellent plasticizer for cellulose nitrate because it produced a more stable and usable product. This product he christened celluloid. One of the initial uses of celluloid was to make dental plates (replacing their unsatisfactory rubber-based predecessors), and in 1870, Hyatt and his brother Isaiah set up the Albany Dental Plate Company. By 1872, the name had been changed to the Celluloid Manufacturing Company. The Hyatt brothers' company was very successful, producing large numbers of saleable goods, including combs and cosmetic boxes.

Spill felt that Hyatt's work on celluloid infringed the British patents, and he entered an eight-year-long litigation battle in the American courts. After an initial favorable verdict for Spill, judgment eventually went in favor of Hyatt, and celluloid became a runaway commercial success. Spill died an embittered man.

How should we assess the contribution of Parkesine and the work of Alexander Parkes to the history of plastics? Parkes was the first to produce a moldable material based on cellulose nitrate, which later developed into what we know as celluloid, abundant in the late 19th and early 20th centuries for hair adornments, decorative items such as dressing-table sets, bags and, perhaps most notable, low-cost, wipe-clean collars and cuffs. His work provided an impetus for what is now a vast industry that creates plastics for every conceivable purpose. Parkes almost certainly deserves his title as the "Father of Plastics."

—Sue Mossman

THE LENOIR GAS ENGINE (c. 1865)

In *The Engineer* of October 26, 1860, there appeared a brief notice with the headline "New Motive Power—Lenoir's Gas Engine." It stated:

A machine placed in a corner of a room, of a size proportionate to the force intended to be developed, the pipe which brings into the manufactory the gas which lights it, a little electric pile on a bracket against the wall, a little water supplied by the water company and atmospheric air, and we have all that constitutes the moving power of M. Lenoir. The machine itself is little more than an ordinary horizontal engine, but instead of it being brought into operation by the expansive force of steam, motion is produced by the combustion of a mixture of gas and atmospheric air; and this combustion is brought about by the action of an electric battery.

The Lenoir gas engine was the first practical internal combustion engine, a forerunner to today's gasoline and diesel engines. Piped supplies of coal gas, principally for lighting, had become widely available in large towns in Britain by the middle of the 19th century, and devices to produce an electric spark had been developed. A compact engine that used gas and an electric spark to ignite it was therefore seen as a major improvement on the steam engine, which needed a boiler, bulky fuel, and a large water supply.

The cannon, of which the first recorded use is believed to have been in the 14th century, may be regarded as the ancestor of the internal combustion engine: the idea of producing mechanical work from the combustion of a rapidly burning fuel in an enclosed space has a long history. Late in the 17th century, the physicist Christiaan Huygens (1629–95), noted for his wave theory of light, made the first gunpowder engine that actually worked. In Huygens' machine, a charge of gunpowder was fired in the bottom of a cylinder in which a piston rested at the top; valves allowed much of the expanding gases to escape. As what remained cooled, a partial vacuum was formed in the cylinder, causing the piston to be forced downward by atmospheric pressure. A weight, attached to the piston by a rope passing over a pulley, was thereby raised. There is clearly a parallel here with the early steam engine: atmospheric pressure acts on a piston in a cylinder in which a partial vacuum has been created by the condensation of steam.

From the 1790's onward, experimenters made engines using various kinds of fuel, including hydrogen and other gases obtained from the distillation of coal, wood or oil. They also tackled the problem of producing continuous operation, of which Huygens' engine was clearly not capable. Some of the engines they developed were direct-acting and others worked on the atmospheric principle. Lenoir's machine was the first to progress beyond the experimental stage and into commercial production. Estimates are that the total number built approached 500; most were made in France, but some were manufactured in England, Germany, and North America. Production ceased around 1869.

Jean-Joseph Étienne Lenoir (1822–1900) was a Belgian who went to Paris at the age of twenty-six to work as a metal enameller. His inventions included a method of electroplating and a railway telegraph system. His gas engine resembles in form and sequence of operation a single-cylinder, double-acting steam engine, in that pressure acts alternately on each side of the piston. Air and gas are drawn into the cylinder during the first part of each stroke. The mixture is then fired by an electric spark, and the expanding hot gases drive the piston to the end of the stroke. As the piston moves to the end of its travel, it forces out, through the exhaust valve at the other end of the cylinder, the spent gases from the previous power stroke. Cooling of the cylinder is assisted by a water jacket. A battery formed from two Bunsen cells provides the ignition via an induction coil and a distributor serving a spark plug at each end of the cylinder.

About one hundred Lenoir engines were built in England by Reading Ironworks Ltd. The engine pictured opposite is nominally of one-half horsepower, and it drove the workshops of the Patent Office Museum, London, from 1865 to 1868. An engine of this size would have cost £65. With regular skilled maintenance, it gave satisfactory service. A contemporary record states that a peak indicated output of about one horsepower (.75 kW) was achieved when it was running at 110 revolutions per minute. In less skilled hands, however, Lenoir's engines were often troublesome, and their performance was soon surpassed by other engines—notably those developed by the German partnership of Otto and Langen. To Lenoir, however, must go the credit for pioneering work on which others were able to build.

— PETER STEPHENS

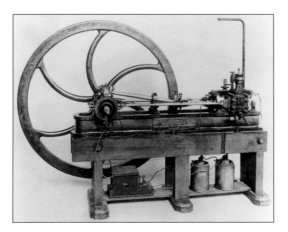

The Lenoir gas engine, 1860

BESSEMER CONVERTER (1865)

The Bessemer converter was developed in 1856 by Henry (later Sir Henry) Bessemer (1813–98) for the production of bulk steel.

Steel is simply iron containing a small amount of carbon (typically one tenth to one and a half percent, depending on the properties required); it is a versatile material suitable for many applications. Today, a wide range of steels and steel alloys can be produced, but prior to Bessemer's invention, steel was available only in small quantities of variable quality, produced by either the cementation or crucible processes. Engineers were restricted to the use of wrought iron, which was relatively soft and weak, or cast iron, which was hard but brittle.

Bessemer is often described as a "professional inventor" because of the variety of products he devised. He was not a metallurgist, but he was trying to find a malleable form of iron that would be tough enough to make shells for a rifled gun barrel he had invented. His small workshop at Baxter House, Paddington, London, was to become the focus of his activities.

After some preliminary experiments, Bessemer realized that if sufficient air could be blown into molten pig iron (which contains about three percent carbon), some of the carbon would combine with the oxygen in the air and be burned. Once the iron was melted and poured into the converter he had created, no further fuel would be required, because sufficient heat would be obtained from the chemical reaction.

In fact, the reaction was extremely violent, and during Bessemer's first experiment no one could approach the converter to turn off the blast, so everyone present had to watch from a respectful distance until the reaction had subsided. Bessemer made some modifications to the design of the converter to contain the burning debris, and in August 1856, he invited the engineer George Rennie (1791–1866) to witness his new process.

Rennie was immediately enthusiastic, and he arranged for Bessemer to present a paper at a meeting of the British Association for the Advancement of Science, in Cheltenham, a few days later. Ironmasters present at this meeting were very skeptical of a paper entitled "On the Manufacture of Malleable Iron and Steel without Fuel," particularly as it was to be presented by an outsider.

Several leading ironmasters did, however, approach Bessemer for patent rights, and they visited Baxter House for demonstrations. Having paid considerable sums of money in royalties, they went away to begin experimenting with what they had purchased. Without exception, they failed—because the steel they produced was brittle and useless. They expressed their annoyance to Bessemer, but neither he nor anyone else could offer a solution. Later, it became evident that the problem was caused by phosphorus in the iron ore. By chance, Bessemer, for his own experiments, had obtained iron made from one of the few sources of low-phosphorus ore in Britain, Blaenavon in South Wales.

Other low-phosphorus iron ore deposits occur in England's Lake District. Here the Barrow Haematite Steel Company began experiments in 1864, using the small converter pictured opposite. The first cast of molten steel was poured in 1865. Having proved successful in trials, the small converter was removed in 1866, and a full-size Bessemer works established. By the next year, the company was the largest Bessemer steelworks in Britain. From his first experiments, Bessemer had now modified the design to a tilting converter, which allowed the vessel to be turned on its side to pour out the molten steel.

Gradually, the Bessemer process became established using low-phosphorus iron ores, but efforts to solve the phosphorus problem (so that other high-phosphorus iron ores could be used) were not a complete success. The eventual solution was devised by two cousins, Sidney Gilchrist Thomas (1850–85) and Percy Carlyle Gilchrist (1851–1935). Thomas had been challenged by a chance remark at an evening class that whoever solved the phosphorus problem would make a fortune. After some preliminary experiments (initially carried out in his bedroom), he enlisted the help of his cousin, a steelworks chemist. They worked out that if the converter was lined with bricks that were chemically basic (a base is able to neutralize acid), the furnace lining would neutralize the acid phosphorus and remove it from the reaction. They had great difficulty in convincing ironmasters that they had succeeded where others had failed, but their experiments did provide a solution, and it was adopted from about 1880.

In the meantime, another furnace had been devised by Sir William Siemens (1823–83), working in Britain, and, independently, by Pierre-Émile Martin (1824–1915) in Sireuil, France. Their "open hearth" furnace could be controlled more easily to get the carbon content of the steel exactly right and could also recycle steel scrap. From the early 1870's, this process developed alongside Bessemer's, but it was also dependent upon the results of the work of Gilchrist and Thomas for its ultimate success.

The availability of large quantities of steel at an acceptable quality and price added a new dimension to engineering. It made possible much larger bridge spans and consequently larger ships, taller buildings, stronger and lighter machinery, and steel rails that paved the way for heavier and faster trains.

In more recent times, bulk steelmaking has relied entirely upon a derivative of the Bessemer process, basic oxygen steelmaking (BOS). A much larger converter is used, and pure oxygen is pumped into the molten metal using a water-cooled lance inserted into the mouth of the vessel. One converter can produce more than 350 tons of steel in forty minutes.

—Sue Cackett

SWAN AND *RAVEN* (1867–70)

In June 1867, the retired engineer William Froude (1810–79), assisted by his son Edmund (1846–1924), began testing a series of model ship-hull shapes in a small creek off the River Dart opposite the town of Dartmouth, in Devon, England. These models, which Froude named *Swan* and *Raven*, had as their principal purpose the establishment of a method whereby a scale model could be used to predict the resistance (or "drag") experienced by a full-sized ship as it moved through water. Such information was necessary to enable naval architects to specify the size of steam engines needed to drive a ship through water at its required speed.

Although a great deal of effort had been expended in the 18th and early 19th centuries in using models to estimate the resistance properties of various hull shapes, all experiments had failed to achieve their purpose. The reason was that nobody had been able to discover how model test results could be reliably scaled up. It was well known that they did not scale up directly in proportion to the scaling factor of the model. The successful application of ship models to ship design therefore needed a law to relate the model's resistance, and the speed at which it was measured, to its full-sized counterpart. The law to which the Froudes were trying to adhere is known as the law of similarity.

To investigate and discover the law of similarity in his experiments, William Froude constructed three sets of *Swan* and *Raven* models. The first were three feet in length, the second six feet, and the third twelve feet, thus introducing a scale factor of two between successive models. Each set of models was then towed simultaneously by a steam launch, their respective resistances being measured by means of a recording dynamometer which also recorded the launch's speed. This procedure was carried out at a number of different speeds, and the resulting speed-resistance graphs for each set of models were plotted for comparison. There was

William Froude

indeed a degree of similarity in the curves for each hull form across the scales, and each plotting demonstrated that above a certain speed the *Raven*, or "sharp" form, showed greater resistance than the *Swan*, or "blunt" form. This was a significant finding, because *Raven* possessed a hull shape that was thought to guarantee a resistance lower than any other shape. The plotted results seemed to demonstrate that this assumption was wrong.

The River Dart experiments with *Swan* and *Raven* were conducted sporadically until early 1868, when Froude communicated the results to the Admiralty. He was able to demonstrate that the resistances between models varied as the cube of the scaling factor, but he was unable to offer a convincing explanation of how speed was related to the resistances. This dilemma was solved during the summer and autumn of 1868 when he developed what later became known as Froude's hypothesis. Two components combined to make up ship resistance: the first was water friction on the surface of the hull, and the second was the differences in pressure over the hull that resulted in the generation of the wave system that accompanied the hull. Since the generation and maintenance of these waves took energy from the ship's engines, they were a form of resistance; he called them "wave resistance." Using this idea and his knowledge of water friction, Froude showed that the two resistance components did not scale in the same way. He further demonstrated that the similarity condition for the wave resistance across scales was met, providing the waves made by each scale of model diverged from the models' bows at the same angle.

The arguments that Froude presented to the Admiralty to justify his ideas were more complex than this explanation might suggest, but they were of sufficient authority to be accepted as sound. On this basis, the Admiralty constructed the world's first ship-model testing tank in 1870 at Chelston Cross, Torquay, adjacent to Froude's house. The tank was so successful that it was soon followed by others constructed by both private ship builders and foreign governments. From the time of its inauguration, the Chelston Cross tank was used routinely to examine the resistance properties of all new hull designs proposed by the Admiralty. It was also employed in fundamental research into the design of propellers as well as a host of other hydrodynamic problems relating to the design of ships and to predicting their performance.

—Tom Wright

HOLMES' LIGHTHOUSE GENERATOR (1867)

One of the greatest breakthroughs in 19th-century science was Michael Faraday's fundamental discovery that an electrical current is produced when a wire cuts through a magnetic field. He discovered this effect, known as magnetoelectric induction, in August 1831. In November of that year, this celebrated British chemist used this principle to construct a disk generator capable of producing a relatively continuous electric current. In so doing, Faraday opened up a completely new field for the development of mechanical devices that transformed energy continuously from a mechanical to an electrical form.

A stream of magnetoelectric generators emerged in quite rapid succession over the next thirty years. Each one was commonly known by its maker's name, and each one was supposed to involve an improvement in efficiency or output over its predecessor.

In about 1850, Florise Nollet (1794–1853), Professor of Physics at the Military School of Brussels, was working on the problem of lighthouse illumination. He intended to produce a brilliant light (limelight) by heating a lump of lime in an oxy-hydrogen flame. To produce the necessary oxygen and hydrogen gases, he built a crude disk armature machine with coils and magnets, intending to break down water by electrolysis. The experiment was not a success. In 1853, Nollet died, and Frederick Hale Holmes (c. 1811–70), an analytical chemist and engineer from London, was invited to continue Nollet's work. Holmes' background is obscure, but he responded and thereafter became associated with the Compagnie de l'Alliance, the company set up to develop Nollet's work.

In Paris, while working on these magnetoelectric generators, Holmes realized that they might provide a continuous current capable of producing a brilliant electric arc that could be used for lighting. By 1855 he had built a number of successful machines, and he had acquired sufficient knowledge, experience and confidence to solicit the Elder Brethren of Trinity House (the English lighthouse authority) to test his magnetoelectric generators and arc lamps for lighthouse illumination. Trinity House sought Michael Faraday's advice and agreed to a trial at Blackwall in 1857 under his direction. The generator was five feet long, five feet wide, and four and a half feet high; it weighed two tons.

The generator proved successful, and, with Faraday's backing, two larger Holmes machines were installed in the South Foreland lighthouse. It was there on the Kent coast, on December 8, 1858, that electric light first became fully utilized in a lighthouse. Faraday's role as consultant for Trinity House was both influential and instrumental in a further Holmes machine being taken into use at Dungeness in June 1862. That machine would run (intermittently) for the next thirteen years.

The Holmes magnetoelectric machine pictured opposite was designed in 1867, and it was one of a pair put into service on January 12, 1871. The location was Souter Point, a new lighthouse erected the year before on the east coast of Great Britain between South Shields and Sunderland. Prior to its installation, this machine was exhibited at the Paris Universal Exhibition of 1867, and afterward, it was erected at Trinity Wharf, Blackwall, for experimental purposes.

Faraday's wholehearted endorsement of Holmes' machines directly influenced Trinity House's receptiveness to electric light—and led to the adoption of Holmes' system in 1857. But by the 1860's, it was no longer clear that Holmes was ahead in the development of magnetoelectric machines. Ironically, it was the appearance of Holmes and this machine at the Paris Universal Exhibition that did much to bring to the attention of the English lighthouse authorities the significant progress made by the French in this field.

From a purely technological standpoint, it is quite remarkable that this dinosaur of a machine ever reached the installation stage at Souter Point as late as 1871. Perhaps Holmes' long association with Trinity House, the latter's innate conservatism, and a sense of loyalty to Faraday (who had died in 1867) combined to ensure that the Souter Point lighthouse was lit by Holmes' 1867 machine. It remained in service there until 1900, but long before then the market had come to be dominated by dynamo-makers Siemens of Germany, Gramme of France, and Brush of America, with Brush providing a range of complete arc-lighting systems for quite different applications.

—David Woodcock

JULIA MARGARET CAMERON'S *IAGO* (1867)

The world's earliest photographs, taken more than a century and a half ago, needed very long exposures. None were portraits, for nobody could keep sufficiently motionless for the minutes, even hours, required. But, as soon as the two prototype systems—the English calotype and the French daguerreotype—reduced exposure times to fewer than sixty seconds, things changed: perhaps as many as ninety-five percent of all the daguerreotypes ever taken were portraits.

During the 1840's, scores of portrait studios were opened, first in London and Paris, then in cities large and small throughout the world. Their owners, more interested in quick profits than aesthetics, rushed their customers through the experience of being photographed. Not surprisingly, most of the resulting pictures have little claim to any artistic merit. There were, of course, exceptions. In Scotland, the likenesses of hundreds of scientists, artists, clerics and leaders of Edinburgh society were captured by the painter David Octavius Hill and his partner Robert Adamson, in carefully posed and lit studies. Taken in the mid-1840's, they are perhaps the first masterpieces in the history of photography.

Twenty years later, a plain, bossy, ambitious and middle-aged English gentlewoman was given a camera by her daughter and son-in-law, in the hope that it would keep her busy while her husband visited his coffee plantations in Ceylon. Their hopes were more than realized: Julia Margaret Cameron (1815–79) became the greatest photographic portraitist of her time.

Mrs. Cameron used Frederick Scott Archer's "wet-collodion" process (*see* **Lewis Carroll's Photographs**). The operation was complex, difficult, even dangerous, especially for a woman who had been looked after by servants all of her life. Not surprisingly, Cameron was very proud of her first portrait. The note she sent to the father of the young girl sitting so still in it (for perhaps a minute or two) gives some idea of the scale of Cameron's problems—and of her triumph in solving them:

Sir John Herschel *by Julia Margaret Cameron, 1867*

"Given to her father by me…My first perfect success in the complete Photograph…This Photograph was taken by me at 1 p.m. Friday Jan. 29th. Printed—Toned—fixed and framed all by me & given as it now is by 8 p.m. the same day."

A photograph needing such a long time, and so much effort, was unlikely to be as stereotyped and dull as a commercial portrait. The struggle to master the chemistry, and to keep the sitter still throughout a lengthy exposure, helped many 19th century amateur portraitists produce far more memorable results than did their professional counterparts.

Cameron lived on the Isle of Wight, a neighbor of Alfred Lord Tennyson, Queen Victoria's Poet Laureate. Through him, she met and photographed many notables of the day, among them Browning, Carlyle, Darwin and Herschel. To Herschel, who

introduced her to the finer points of photographic technique, and whom she called her "teacher and high priest," she gave one of the finest albums of her pictures. Among her portraits of Tennyson and other distinguished Victorians, "the peasantry of our island," and her own family, is the unique *Iago: Study from an Italian*. No other example is known, and it may be that, after making one print, she accidentally broke the negative; it often happened.

Iago seems to be the only example of Cameron's using a professional model. Angelo Colarossi was one of a group of London models, mostly men, who satisfied Victorian artists' demands for Italians and their "feeling for classical stance." One of the painters who employed him was Cameron's friend George Frederic Watts, and in an early version of Watts' *The Prodigal Son*, Colarossi can be seen with the bristly beginnings of a beard. The resemblance to *Iago* is unmistakable.

Perhaps Watts asked Cameron to take the photograph for his reference. Or perhaps she visited him while Colarossi was posing in his studio. She was certainly fond of Italian subjects; there are two others in the Herschel album.

Mrs. Cameron's life-size close-ups have a powerful directness unequalled in portraiture. Compared with the tiny *carte de visite* portraits then in vogue, they are almost overwhelming in scale and personality. When she photographed great artists and scientists, she was consciously "recording faithfully the greatness of the inner as well as the features of the outer man," and she captured the personality of her Italian model just as effectively. Was she aware of something evil as he stood absolutely still before her bulky camera? Or did she sense it as the murky image slowly emerged in the darkroom? Only then, perhaps, considering the complex villain in Shakespeare's *Othello*, did she decide to call this compelling picture *Iago*.

—COLIN FORD

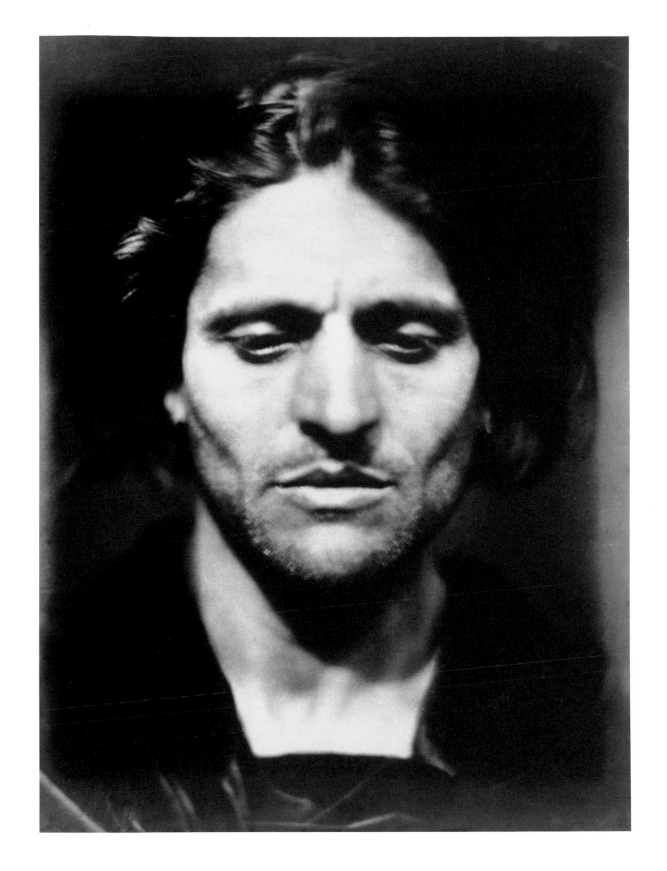

Until the first years of the 21st century, cathode-ray tubes were widely used: they formed the screens in television sets, computer monitors, radar sets, and measuring instruments.

The characteristic glow produced by passing an electric current through air at low pressure has been known since the early 18th century. In 1855, Heinrich Geissler (1814–79), a glassblower in Bonn, devised a new pump that achieved a much higher vacuum. Geissler made the colorful Geissler tubes, which delighted lecture audiences, but he also made apparatus used by Julius Plücker (1801–68) at the University of Bonn. Plücker noticed that at very low gas pressures the electrical glow discharge disappeared, and was replaced by a green fluorescence from the glass at one end of the discharge tube. This fluorescence was caused by rays emanating from the negative electrode, or cathode.

William Crookes (1832–1919), an experimental chemist and physicist, came to the phenomenon by a different route. In 1861, he discovered a new element, which he called thallium. Over the next ten years he measured its atomic weight very carefully. Many samples had to be weighed in a vacuum, and Crookes noticed certain inconsistencies in the weights: warm bodies seemed slightly lighter than cooler ones. He set out to examine the effect, partly out of scientific curiosity and partly because he supposed it might be connected with psychic and paranormal phenomena, which fascinated him.

This work led Crookes to devise, in 1875, the "light-mill," or radiometer. Crookes' radiometer consists of a vane with four arms, each blackened on one side, balanced on a point in the center of a glass bulb from which most of the air has been removed. When light or radiant heat falls on the vane, it rotates, the blackened faces moving away from the radiation. This effect depends on the blackened faces of the vane absorbing more radiation than the other faces and thus becoming slightly warmer.

In other experiments, Crookes passed an electric current through a radiometer, which produced the familiar glow from the residual gas in the bulb. When the vane was made negative, there was a dark space

William Crookes, cartoon by Spy from Vanity Fair, *1903. Opposite: Maltese-cross tube and radiometer*

between it and the glow. The dark space was larger on the blackened faces of the vane. Believing that it could help his understanding of the radiometer, Crookes began a new series of researches on "Crookes dark space." As the gas pressure was reduced, the dark space increased until it filled the whole of the bulb, and the fluorescence seen by Plücker appeared. Crookes investigated the cathode rays that caused this fluorescence.

Crookes reported this work to the Royal Society in the Bakerian Lecture of 1879. Subsequently, he was asked to lecture at the Royal Institution and to the British Association for the Advancement of Science. It was the climax of his career. His demonstrations aroused enormous interest. Cathode rays caused minerals such as ruby to phosphoresce, as

well as the glass of the discharge tube. The stream of cathode rays could be deflected by a magnet, showing that they were electrically charged, but otherwise they traveled in straight lines. This phenomenon was shown by using an obstacle, in the shape of a Maltese cross, to cast a shadow on the fluorescence at the end of the discharge tube. More than a century later, a form of the Maltese-cross tube is still used to demonstrate the properties of cathode rays, which are now known to be beams of subatomic particles called electrons.

The radiometer has never been more than a popular scientific toy, but Crookes' work on cathode rays was seminal. In 1895, Wilhelm Conrad Röntgen (1845–1923) noticed that when cathode rays struck the end of a discharge tube, rays of a different kind were emitted, capable of penetrating matter. He called them x-rays, and their significance can hardly be overstated. Crookes was disappointed not to have discovered x-rays himself.

In 1897, Joseph John Thomson (1856–1940) made measurements that suggested to him that cathode rays might be streams of particles, each carrying a single unit of charge, much lighter in weight than any atom. This event is regarded as the discovery of the electron, and much of modern electronics depends upon it (*see* **Thomson's Cathode-Ray Tube**).

Also in 1897, Ferdinand Braun (1850–1918) used a form of Crookes' tube as a measuring instrument. It was the direct antecedent of today's cathode-ray tube, though it required the discovery of the thermionic emission of electrons from a heated filament before it could take its final form.

From about 1881, Crookes' former assistant, Charles Gimingham, used his skills in glassblowing and vacuum technique to help Joseph Swan put his electric filament lamp into production, thus helping to usher in the age of electric lighting. The electric lamp in turn led to numerous advances in electronics. It is safe to say that the ramifications of Crookes' work are enormous.

—Neil Brown

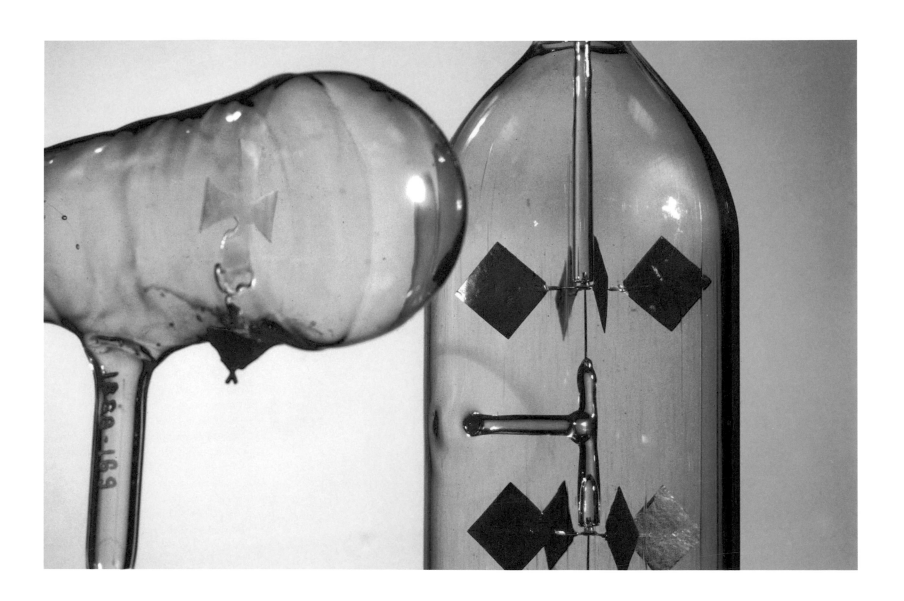

In 1981, the American communications corporation AT&T came up with the slogan "Reach out and touch someone." The idea was simply to provoke customers to make more use of the telephone, but the phrase evoked echoes of an incident that had occurred in England 103 years earlier.

Alexander Graham Bell (1847–1922), inventor of the telephone and effectively the founder of AT&T, came from a Scottish family obsessed with speech and hearing. His grandfather, father and uncle all taught speech and elocution. His father married a deaf girl who was to become Bell's mother. Bell, in turn, became a speech practitioner in Boston, and he married one of his clients, a young woman who was profoundly deaf.

It was possibly this familiarity with the problems of the deaf that led Bell into his royal *faux pas* of 1878. Bell had been invited to demonstrate his telephone to Queen Victoria at her residence, Osborne House on the Isle of Wight. On the evening of January 14th, he made contact with nearby Osborne Cottage and also with Cowes, Southampton, and London. Among the telephones used that evening was the instrument pictured opposite.

At some crucial point of the demonstration, the Queen turned away from her telephone. Bell, accustomed to attracting the attention of the deaf, leaned over and touched her arm, becoming the first commoner in years to lay hands on royalty without permission. Queen Victoria may or may not have been amused, but his action certainly gave Bell unlooked-for publicity in court circles.

The walnut-and-ivory decoration lavished on this crude telephone by instrument makers Julius Sax and Company indicates how eager Bell was to impress. By 1878, the telephone, introduced in America in 1876, was familiar to British technical specialists as a result of the lectures and demonstrations given by the British physicist Lord Kelvin (1824–1907) and Sir William Preece (1834–1913), Engineer-in-Chief of the Post Office. But the big push toward commercial success was just beginning, and Bell needed all the patronage he could get. He employed one of the world's first public relations advisers, the American journalist Kate Field, who, as well as being on hand to sing on the telephone to Queen Victoria from Osborne Cottage, kept newspapers on both sides of the Atlantic amply supplied with material promoting the telephone.

Field's accounts paint the Osborne demonstrations as marvelous entertainment, with singing and laughter heard distinctly over the wires, but the demonstration was in reality only a partial success. The line from Cowes failed, and the line from London was so bad that Bell did not attempt to transmit speech, confining himself to the tones of an organ. Whatever Field said, there were clearly many practical problems still to be overcome.

One of them was the question of the transmitter. The Bell telephone used identical devices for transmitter and receiver, each consisting of an iron diaphragm, a magnet, and a coil of wire. Sound waves in the transmitter made the diaphragm move, producing electric currents in the coil and sending electric waves down the line to the receiver. With the receiver, the effect was reversed, with currents in the coil moving the iron diaphragm, recreating the original sound. The system lacked amplification, however, making its range frustratingly limited. It was Thomas Edison (1847–1931) who, a month after the Osborne demonstration, patented the amplifying carbon microphone, which provided a partial solution to the problem of transmitting telephone signals over long distances.

Of the other instruments that survive from the Osborne demonstration, there are three in the Smithsonian Institution in Washington, D.C., and one at the AT&T Archives in Warren, New Jersey—although all of these instruments lack the wooden switchplate provided by Sax. This one is labeled "Osborne Cottage," and it was probably used not by Queen Victoria but by her attendant Sir Thomas Myddelton-Biddulph and his wife. They are reported to have conversed with the Queen, while Kate Field, after her contribution, had to make do with Queen Victoria's thanks relayed by the Duke of Connaught. Following the demonstration, the Queen, despite having described the telephone in her journal as "rather faint," was sufficiently impressed to ask Bell if he would sell two of the instruments left at Osborne. This sale appears never to have happened.

The telephone and other electronic communication devices are taken so much for granted now that it is difficult to recapture the excitement that Bell's tinny simulations of human speech must have elicited at the time. Still less can one imagine the social confusion created by this unfamiliar visitor who could enter a house without being presented at the front door. The modern smartphones, which are personal, ubiquitous—and perhaps even intrusive—do, it can be argued, derive from Bell's fundamental invention. It can certainly be said, however, that the current fast-moving era of digital and wireless communication does enable one to appreciate the impact of Bell's invention on those who, like Queen Victoria at Osborne, were experiencing it when it was brand new.

—ROGER BRIDGMAN

Pair of early Bell telephones, 1878

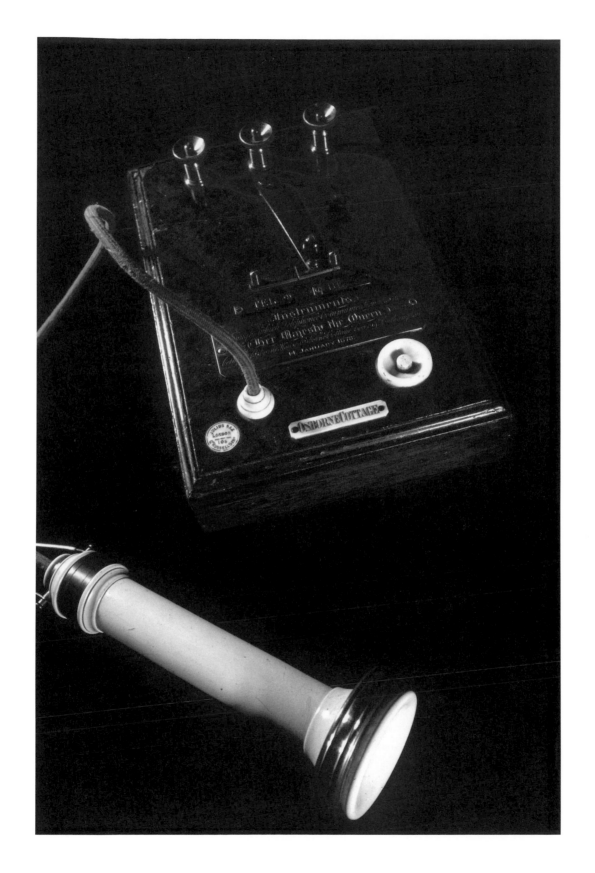

WIMSHURST'S ELECTROSTATIC MACHINE (1882)

Victorian culture saw science as an endless source of wonder and inspiration. Social soirées often took the form of demonstration evenings at which the middle classes would learn of the latest ingenious devices. Ordinary people were also interested in learning as much as they could about scientific developments. Electricity was a particular source of fascination to both groups.

Static electricity—caused by a build-up of electrically charged particles—was the first form of electricity known to man: the word electricity derives from the Greek word for amber, a substance that readily develops an electrostatic charge when rubbed.

In Victorian England, amber was replaced by glass or sealing wax, which, when rubbed with silk, could demonstrate the basic principles of static electricity by attracting small pieces of paper or feathers. Many Victorian devices or toys, such as the electric puppet or the electric cannon and wheel, were used to demonstrate the mysterious qualities of this kind of electricity.

Beginning in 1860, the need for reliable high-voltage machines for applications in telegraphy and laboratory experiments stimulated experiment.

Friction, as a method of producing high-voltage static electricity, was unreliable. The electrophorus, invented by Volta in 1775, used another method called induction. The development of induction machines in continental Europe and Britain culminated in a machine designed by James Wimshurst (1832–1903) in 1882. Known simply as Wimshurst's electrostatic machine, it completely revolutionized the science of static electricity. It was the first machine capable of producing high-voltage static electricity that was not affected by atmospheric humidity.

When the relative humidity is high, electrostatic phenomena, such as the electric charge created when one rubs a balloon against a wool sweater

The "electric bath" from the Journal of Electro-Therapeutics, May 1893

or walks across a synthetic rug, nearly disappears. The reason for this is that layers of moisture on the surfaces allow electrical charges to "drain off" to earth and not accumulate. Likewise, many electrostatic machines could not accumulate a charge and would not produce electricity in humid conditions. Wimshurst's machine was the first to work in almost all atmospheric conditions. It was also highly reliable and easy to use.

Wimshurst attributed this remarkable success to the fact that he had not followed what he called the "orthodox views of electrical science" but had worked with complete originality of thought. And it is true that his machine has not been improved upon significantly since its creation.

The two fourteen-and-a-half-inch diameter disks rotated in opposite directions, with positive charge building up on one disk and negative charge on the other. When discharged, they produced sparks that could jump a distance of four and a half inches. Wimshurst's largest machine had disks of seven feet in diameter, and it produced sparks that could jump a distance of two and a half feet, corresponding to several hundred thousand volts.

At the turn of the century electrostatic machines such as that of Wimshurst were used to generate high voltages to power x-ray tubes. Another common application was in electrotherapy. A patient would be charged up in an electric bath or subjected to electric shocks. These treatments, although considered very efficacious for insomnia, rheumatism, sprains, bruises and even constipation, were soon replaced by less dangerous and more effective methods. Most conditions that were once treated by electrotherapy are now treated by chemical means.

Today, electrostatics is important because of its various uses. Electrostatic precipitators are used in industry to remove contaminating particles from gases; they are used in domestic air conditioners to remove dust. Photocopiers and laser printers employ the principles of electrostatics, as do spray painting and crop-spraying devices. Electrostatics does involve dangers. Precautions must be taken to prevent the build-up of electric charge on aircraft, electronic equipment and in situations involving flammable substances.

James Wimshurst was of the amateur scientist breed, although he was a marine surveyor by profession. His obituary states that what he did "he did for the love of work and advancement of science—all for love and nothing for reward."

—Peter Bailes

THE FIRST TURBO-GENERATOR (1884)

Look here, you fellows, I have an engine that is going to run twenty times faster than any engine today. —The Hon. C. A. Parsons

Charles Algernon Parsons (1854–1931) was the youngest of the three sons of the Earl of Rosse, an accomplished Victorian astronomer (*see* **The Rosse Mirror**). The family seat at Birr Castle was a focus for scientific thought and discussion among those who came to use the telescope there. Astronomers employed by the Earl to operate his observatory by night were additionally required to tutor his children by day. Charles and his brothers enjoyed regular access to the estate's workshops, and they grew up in an atmosphere of lively intellectual inquiry allied to practical engineering skills. These skills were nurtured in Charles' case by two years at Trinity College, Dublin, followed by a four-year degree course in mathematics at Cambridge. Much of his time as an undergraduate was devoted to mathematical experiments. He was particularly preoccupied by the limitations on speed and size that then confronted the builders of reciprocating steam engines, as they sought new ways of exploiting the increase in boiler pressure then becoming available. Remarkably, for a young man of his privileged background, C.A. Parsons enrolled in 1877 as a premium apprentice at Armstrong's works at Elswick, where he perfected his mechanical skills.

In 1881, he obtained an appointment with Kitson's, engineers in Leeds. Through this period he wrestled with the engineering problems of steam-turbine design. The main problem was far from novel: James Watt's steam wheel was one of many hundreds of attempts to harness the expansive power of steam to produce rotative power directly, without the intervention of a piston, which wasted energy. It was known that to obtain efficiency, the vanes of a steam turbine should travel at about half the speed of the steam passing over them. Steam escaping to the atmosphere at a pressure of 200 pounds per square inch can achieve a velocity of 4,000 feet per second. Parsons realized that he could harness such high velocity if he divided the

A portable Parsons turbo-generator, 1885. A similar generator was used to light open-air ice skating in January 1886, as a novel way of raising money for a Newcastle hospital (NEI Parsons)

pressure drop into several stages, so that the reduction in the steam's velocity was gradual, and work could be extracted from its expansion at each stage.

Parsons settled on fifteen progressive stages of expansion for his first practical steam turbine, which he completed in 1884. Sets of fixed vanes attached to the turbine casing separated the rings of moving vanes attached to the turbine shaft, directing the steam onto them and helping to accelerate the flow of steam. The machine ran at 18,000 revolutions per minute (rpm), unprecedented at the time for a steam engine. Parsons recognized the need to limit the diameter of the rotor so that centrifugal forces remained manageable and the engine remained balanced. A neat solution to the problem of end-pressure on the axial turbine was achieved by his admitting the steam at the midpoint of the turbine shaft and dividing the flow into two equal streams, the axial loadings from which balanced each other, so that no thrust bearing was necessary. The turbine shaft had to be slender to minimize the external diameter of, and therefore the centrifugal force on, the rotors, and it had to be long to accommodate a total of thirty sets of rotors and guide rings—due to the fifteen stages of expansion on either side of the steam inlet.

Such a configuration posed problems of trans-

verse vibration as the shaft was accelerated through a series of critical speed ranges corresponding to the natural aural frequencies of the shaft when struck. Once through these critical speed ranges, the "whirling" phenomenon abated, and acceleration could safely continue. Parsons saw that the whirling would be exacerbated if the shaft bearings sought to restrain the shaft altogether, but it could be lessened if the bearings permitted a degree of controlled deflection through the critical speed ranges.

Elegance in engineering is often characterized by simplicity, and Parsons' answer to the whirling problem exemplifies this truism. In the first 1884 turbine, the bearing journals at each end of the rotor shaft have about 1/32 inch of latitude. Each is supported by a large number of washers alternately fitting the bush and the casing. A spring compresses each assembly so that relative movement between adjacent washers is inhibited. Each bearing can align itself to the direction taken by the revolving shaft, but sudden and destructive lateral displacement of the shaft is controlled by the friction between the washers.

Parsons ran his turbine for extended periods at 18,000 rpm, and proved the adequacy of the balancing and lubrication arrangements. To accommodate any whirling of the shaft, he ensured that clearances between the rotor and the turbine casing were generous, and steam consumption was consequently prodigious. Because it demonstrated that sustained high-speed rotation was safely obtainable from a steam engine, the Parsons turbine enabled generators to be greatly simplified and the rotating armature made smaller now that it could be rotated so much faster than before.

The turbo-generator patented by C.A. Parsons in 1884 represents the most important single development in the history of electric-power generation. Today, much of the electricity consumed in the United Kingdom comes from steam turbines, using exactly the same principles as those demonstrated by Parsons in 1884.

—JOHN ROBINSON

ROVER SAFETY BICYCLE (1885)

It is not always appreciated that the design of the present-day bicycle has remained much the same as that which originated in the 1880's. From the early 1870's to the late 1880's, the high bicycle, Ordinary, or Penny-Farthing bicycle as it came to be known, was extremely popular, and its development led to the formation of numerous cycling clubs. By 1882 there were almost 350 such clubs in the British provinces and 184 in Greater London. Besides club members, many others took to the rough roads on weekends, exploring countryside previously impossible to reach in a day.

The design of the Ordinary had many faults. It was unstable, because the cyclist was almost directly over the center of the large front wheel (which was considerably larger than the back wheel), and there was the danger of the rider "taking a header" if, for example, the front wheel hit a stone. The bicycle was difficult to mount and dismount, and the front wheel was both driven and steered at the same time, which tired the rider's arms. The larger the diameter of the front wheel, the faster the speeds achieved: for every turn of the pedals, a greater distance was covered. The only factor limiting the diameter of the front driving wheel was the length of the rider's legs.

The availability of better materials, and the improved technology of components such as chains, led to a search for a safer design. When introduced, the first important new design was so much lower and more stable than its predecessor (the Ordinary) that it became known as the Safety bicycle.

There followed a number of experimental designs, some of which were produced commercially. Though the front wheel became smaller, it remained larger than the rear wheel, and therefore the bicycle still resembled the Ordinary. One of these bicycles was the Bicyclette, patented by Henry Lawson in 1879, which had its chain drive to the rear wheel.

In 1885, many rear-driven Safety bicycles were shown in public for the first time. Among them was the Rover Safety bicycle designed by John Kemp

A period advertisement for the Rover safety bicycle, 1888

Starley (1854–1901), which was exhibited at the Stanley Show in London from January 28th to February 3rd, 1885. J.K. Starley worked as a cycle mechanic until 1878 when he started his own business in Coventry with William Sutton. In November 1896, the Rover Cycle Company Ltd. was formed. From this organization evolved the Rover Company Ltd., which produced many successful models of bicycles, motorcycles, and cars. Cycle production ceased in 1926.

The Rover Safety bicycle pictured opposite is one of Starley's improved designs made in late 1885; it may be regarded as the prototype that set the trend for future technical development and commercial production. The essential advantage of this design is the diamond-shaped frame, which provides for structural strength and stiffness combined with low weight and compactness. Smaller-diameter wheels of almost equal size ensured the rider a low-riding position for greater stability, and a geared-up chain drive to the rear wheel allowed for more efficient pedaling. In 1885, a one-hundred-mile road race was organized for riders of Rover bicycles, in which the existing fifty- and one-hundred-mile records were broken.

Although the Rover Safety bicycle marked a turning point in the development of the bicycle, there was still much more to be done before the cycle in its classic form evolved. The Rover's steering was direct, and the steering head sloped, but the forks were straight, not curved. The frame was not yet triangulated, as, for example, there is no tube from the saddle to the crank bracket. Pneumatic tires were introduced in 1888, and they were in general use a few years later, together with gears and better brakes. The Rover weighed thirty-seven pounds, compared to about twenty-eight for modern touring bicycles and nineteen pounds for road-racing bicycles.

It was J.K. Starley's Rover Safety bicycle that changed the course of cycle development, as by the early 1890's very few Ordinary cycles appeared in manufacturers' catalogues; it was then that the derogatory name of Penny-Farthing was first used to describe the Ordinary bicycles. Most of the bicycles produced since the 1890's have been based on Starley's original Rover design. As Starley said in a paper presented at the Society of Arts in 1898, "… my aim was not only to make a safety bicycle, but to produce a machine which should be the true *Evolution of the Cycle*, and the fact that so little change has been made in the essential positions, which were established by me in 1885, prove that I was not wrong in the cardinal points to be embodied to this end."

—Francesca Riccini

THE KODAK CAMERA (1888)

In the late 1880's, there appeared a camera that was to change the course of photography. Popular photography can properly be said to have started in 1888 with the introduction of the Kodak.

The Kodak camera was the invention of an American, George Eastman (1854–1932). Advertised as "the smallest, lightest and simplest of all Detective cameras" ("detective" was a popular term in the 1880's for handheld cameras), it was a simple wooden box six-and-a-half-inches long, three-and-three-quarter-inches high, and three-and-a-quarter-inches wide. It was small and light enough to be held in the photographer's hands while it was in use.

Even during the difficult years leading up to the Eastman Kodak Company's 2012 bankruptcy filing, the name "Kodak" was synonymous with popular photography; indeed, for years to come, "Kodak" will likely remain a household word all over the world. It was the success of the original Kodak camera that laid the foundations for Eastman to build an enormous international business empire. He chose the name for his new camera with great care: "The letter 'K' has been a favorite with me—it seems a strong, incisive sort of letter. It became a question of trying out a great number of combinations of letters that made words starting and ending with 'K.' The word 'Kodak' is the result." Eastman later explained to the British Patent Office: "This is not a foreign name or word; it was constructed by me to serve a definite purpose. It has the following merits as a trade-mark word: first, it is short; second, it is not capable of mispronunciation; third, it does not resemble anything in the art and cannot be associated with anything in the art."

Taking a photograph with the Kodak camera was very easy, requiring only three simple actions: turning the key (to wind the film); pulling the string (to set the shutter); and pressing the button (to release the shutter and make the exposure). It was, in many respects, the forerunner of the point-and-shoot cameras of the pre-digital-photography age. The lens fitted to the Kodak had a considerable depth of field; objects were in focus from as close as four feet, and it gave a wide (sixty degree) angle of view. That meant that no viewfinder was needed; the camera was simply pointed at the subject to be photographed. Poor definition at the edge of the image area, however, meant that a circular mask had to be used in the camera, placed in front of the film. This mask accounts for the distinctive round (two-and-a-half-inch-diameter) photographs that the Kodak camera produced.

Ingenious, compact and simple to use though it was, the Kodak camera did not embody any revolutionary technological innovations. It was not the first handheld camera, nor was it the first camera to be made solely for roll film. The true significance of the camera, that which makes it a landmark in the history of photography, is that it was the first stage in a complete system of amateur photography. In Eastman's own words: "The Kodak camera renders possible the Kodak system, whereby the mere mechanical act of taking a picture, which anyone can perform, is divorced from all the chemical manipulations of preparing and finishing pictures, which only experts can perform…We furnish any-

Small children playing at the waterside, c. 1890's.
Photograph taken with a Kodak No. 1 camera

body, man, woman, or child, who has sufficient intelligence to point a box straight and press a button…with an instrument which altogether removes from the practice of photography the necessity for exceptional facilities, or in fact any special knowledge of the art."

The Kodak camera was sold already loaded with sufficient film to take one hundred photographs. After the film had been exposed, the photographer mailed the entire camera to the factory, where it was unloaded and the film developed and printed. The camera, reloaded with fresh film, was then returned to its owner, together with the negatives and a set of prints. Previously, photographers had no choice but to do their own developing and printing, which required a darkroom and the knowledge and skill to perform complex chemical manipulations. These factors, more than any others, had delayed the popularization of photography. With the Kodak system, Eastman had not only removed this barrier, but had founded the photographic developing and printing industry. To promote the Kodak camera, Eastman devised the brilliantly simple sales slogan that summed up his new system: "You press the button, we do the rest."

The new convenience did not, however, come cheap. In the United States, the Kodak camera sold for twenty-five dollars; in Britain, it sold for five guineas (£5.25). In the U.S., the developing and printing service cost a further ten dollars; in the U.K., it was two guineas (£2.10). In 1888, a little over eight dollars was a week's wages for many workers; it was one pound in Britain. Through a combination of pioneering mass-production methods and imaginative marketing techniques, Eastman was able to continue the successful process of turning photography into a truly popular pastime. Improved versions of the Kodak camera were produced, and costs were greatly reduced. In 1900, the one-dollar "Brownie" camera was introduced; it sold for five shillings (25p) in Britain. Then, for the first time, the pleasures of photography had been brought within the reach of practically everyone.

—COLIN HARDING

EARLY CINE-CAMERAS (c. 1888)

Louis Aimé Augustin Le Prince and the brothers Louis and Auguste Lumière have all been credited with the invention of cinema. But it is rare that individuals are solely responsible for any scientific or technological invention. The single-lens camera built by Le Prince circa 1888 and the Cinématographe developed by the Lumière brothers in 1895 illustrate this point very well.

Cinema can be described as the projection of moving photographic pictures. Its evolution was dependent on a handful of technical principles. In 1832, Joseph Antoine Ferdinand Plateau (1801–83) constructed a device that created the illusion of movement through the successive presentation of still images each one of which showed phases of that movement. Photography, the permanent record of optically-formed images on light-sensitive material, was perfected simultaneously by Louis Daguerre (1787–1851) and William Henry Fox Talbot (1800–77) in 1839 (see **Talbot's *Latticed Window*** and **Early Daguerreotypes of Italy**). The technical principles for cinematography were essentially understood by that date.

By the 1870's, a number of people were experimenting with the recording and analysis of movement using photographic techniques. Photographic emulsions allowing exposures as short as one thousandth of a second were available—but only on glass plates. By the mid-1880's, the key issues were the need for long strips of pictures and a method of moving the strip intermittently at a fast enough rate to record movement smoothly—about sixteen pictures per second. In 1885, George Eastman (1854–1932) introduced a paper-based roll film (see **The Kodak Camera**).

The Le Prince single-lens camera made use of Eastman's paper-roll film to record a sequence of images. It is claimed that in October 1888, his camera and Eastman's film were used to take twenty consecutive pictures of Leeds Bridge at a rate of about sixteen pictures per second. The camera and the frames still exist, although there is no definite proof of their date. But Le Prince (1842–90?), a French showman engineer and inventor, applied for an English patent on January 10, 1888, and, in his application, he describes the principles of cinematography.

October 1888 was also the month in which French physiologist Étienne-Jules Marey (1830–1904) gave a presentation to the Académie des Sciences in Paris: he showed a series of pictures made at a rate of twenty per second on a roll of Eastman paper film. The chronology is further complicated: on October 17, 1888, Thomas Alva Edison (1847–1931) filed a caveat with the U.S. Patent and Trademark Office that described a picture version of the phonograph.

Le Prince disappeared in mysterious circumstances in 1890. One theory suggested that Edison had him murdered, although recently discovered papers suggest that Le Prince was worried by large

Above: Lumière Cinématographe No. 8 from 1896.
Opposite: Le Prince single-lens camera c. 1888
(the top lens is a viewfinder)

debts, and may have taken his own life—a less dramatic but more plausible explanation. It is possible that his single-lens camera was the first mechanical expression of cinematography, but it may also have been faked by Le Prince's son Adolphe, who, in 1898, testified against Edison's claim to be the inventor of moving pictures.

It was, however, indisputably Edison who introduced moving pictures to the general public when the first Kinetoscope parlor opened in New York City in 1894. The Kinetoscope was a coin-operated machine that offered a "show" lasting about twenty seconds for a single viewer—the original peepshow.

The brothers Louis (1864–1948) and Auguste (1862–1954) Lumière were the most successful photographic plate manufacturers in France. They first saw a Kinetoscope in the summer of 1894. Impressed by the demonstration but put off by the high prices demanded by Edison's agents, they decided to develop their own product. In February 1895, they patented a combined camera and projector that used an intermittent claw derived from the mechanism used in sewing machines to move the cloth (see **Elias Howe's Sewing Machine**).

The apparatus was called the Cinématographe, and the brothers offered the first public presentation of their invention at the Société d'Encouragement pour l'Industrie Nationale in Paris on March 22, 1895. This demonstration was a one-minute film of workers leaving the Lumière factory in Lyons. Encouraged by its reception, the brothers made more films, and on December 28, 1895, for the first time, an audience paid to see moving photographic pictures, which were projected in the basement of the Grand Café in Paris.

The Lumières commissioned an initial production run of 200 Cinématographes from the instrument maker Jules Carpentier. The Cinématographe pictured here is No. 8 from that run. It was purchased in Paris in 1896 and was immediately exported to Paraguay where it stayed in the same family until it was sold in 1990.

—ROD VARLEY

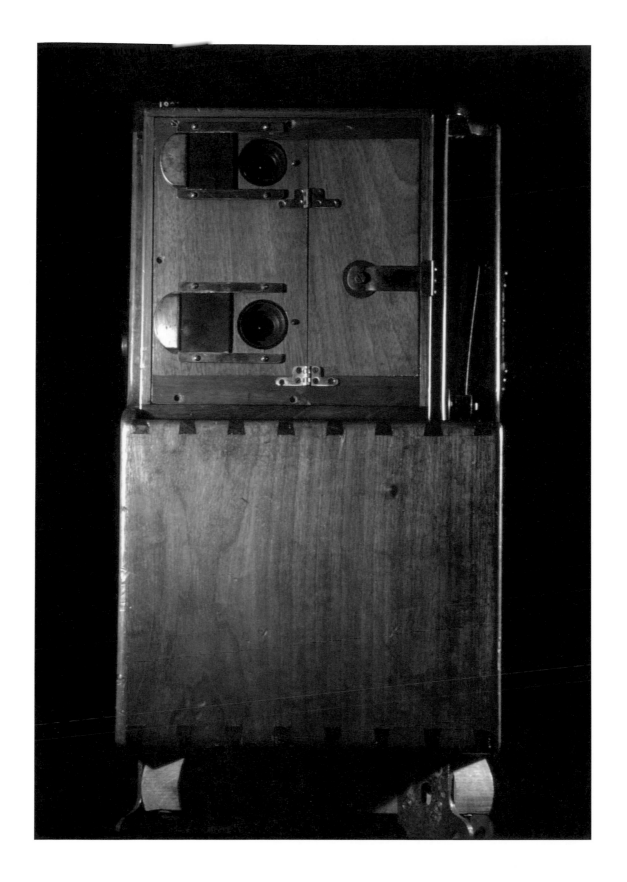

LOCOMOTIVE FROM THE FIRST LONDON "TUBE" RAILWAY (1890)

When it opened in December 1890, the City & South London Railway was the world's first deep-level underground railway. It was electrically powered, and it involved defined stations and a proper system of signaling. It represented an enormous advance both in tunneling and in electrical engineering, but paradoxically it was not designed with electricity in mind. The design of the original locomotive has perhaps as much historical significance to electric railways as Stephenson's *Rocket* has for the development of steam railways sixty years earlier (*see* **Stephenson's Rocket**).

In the mid-19th century, cities throughout the world were experiencing traffic problems little different from those of today, except in scale. In 1863, London pioneered the urban underground railway as an answer to congestion. Engineering constraints meant that the tunnels were constructed by "cut and cover"—digging a deep trench that was then lined with bricks, covered over, and the top surface restored. This method involved extensive demolition of buildings as well as disruption of traffic and essential services; despite these disadvantages, many miles of subterranean railway were built this way in London. The ability of such underground railways to ease congestion, however, was only partial, not least because of the necessity to use steam locomotives, creating a severe ventilation problem.

In the 1860's, an eminent engineer, P.W. Barlow (1809–85), assisted by his pupil, J.H. Greathead (1844–96), developed a shield for tunneling through the London clay at a sufficient depth to avoid existing service installations and building foundations, while causing minimal surface disturbance. With the opening of a short tunnel under the Thames in 1870—the Tower Subway—deep-level underground railways became a practical possibility. Despite many proposals for such railways, by the mid-1880's, only one such scheme had succeeded in getting beyond the planning stage. That was the City of London & Southwark Subway, for which the plan was to haul trains using a continuously moving cable driven by a stationary steam engine. With Greathead as engineer, construction

of the tunnels began in 1886. In 1887, an extension south to Stockwell was agreed to, bringing the route's length to about three and a half miles.

The failure early in 1888 of the company supplying the haulage-cable equipment forced the developers to examine other possible means of traction. Electricity was still a comparative novelty, and in London, there were only a small number of power stations, each supplying electric light to a few local customers. Some lightweight electric railways existed, but nothing on a large scale had yet been tried. It was therefore a daring decision by the CL&SS to opt for electric traction. The builders at first intended to place the motors under the carriages themselves, but in January 1889, the company accepted a proposal from the Salford firm of electrical engineers, Mather & Platt, to use locomotives instead. They were also to supply the other electrical equipment for the line, including the power station. Two prototype locomotives were built, followed by twelve others.

The first prototype was designed by Mather & Platt's chief engineer, Dr. Edward Hopkinson (1859–1922), in conjunction with his brother Dr. John Hopkinson (1849–98), who was retained by the firm as a consultant. It involved a shorter cab than the others. In December 1890, the City

King William Street Station, City & South London Railway, 1890

& South London Railway (formerly the City of London & Southwark Subway), started service using these fourteen locomotives, together with thirty carriages built by the Ashbury Carriage & Iron Co. Ltd. For the first year or so, the locomotives were run in the livery in which they had been delivered, an unlined reddish-brown. The C&SLR then devised a standard color scheme of black with yellow/orange panels and bright yellow lining-out, and the locomotives were converted to this color scheme as soon as they required repainting.

After an unsteady start, traffic on the line did increase, and the line was extended in both directions. (It now forms part of the London Underground's Northern Line.) This expansion required new locomotives to be built, which contained improvements in the electrical equipment, and which could haul longer and heavier cars. Numbers 1 and 2 locomotives were withdrawn from service around 1901, and Numbers 13 and 14 followed in about 1907. The withdrawn engines were stored at the depot at Stockwell, and were cannibalized for parts.

The locomotive pictured opposite was acquired when the C&SLR was rebuilt from 1922 to 1924, and the original trains were scrapped. Its identity had, however, been forgotten: it bore the makers' plates from Number 1, and it was painted in reddish-brown. It was displayed in this guise for more than sixty years until it was finally decided to restore the more representative livery and the correct identity, which surviving records suggest was almost certainly Number 13. The restoration work was carried out in 1990.

Locomotives of essentially this design lasted in service for more than thirty years: a remarkable record, considering that there had been nothing quite like them before. Number 13, in its authentic bright colors, perhaps only now properly commemorates the pioneering achievement of the C&SLR's creators.

—JOHN LIFFEN

PARSONS' MARINE STEAM TURBINE (1894)

By the start of the last decade of the 19th century, the multi-cylindered steam expansion engine had become the most important apparatus used to propel ships. The principle of expanding steam driving a piston to do useful work, by means of passing that steam through three or more cylinders of ever-increasing diameter and lower pressure, had reached the stage at which the physical limitations of the process were apparent. In a period that saw Britain's proportion of world merchant-ship tonnage climb to fifty percent, there were increasing demands for more efficiency and more power to drive bigger and faster naval and merchant ships. This demand created an important market for improved kinds of marine propulsion.

One candidate for this market was the steam turbine, which relied for its operation on steam passing through two sets of blades, or winglets, one set being fixed to the static engine casing and the other to a disk attached to the shaft of the machine. The effect of passing steam through this configuration was to make the disk—and thus the shaft—rotate. By repeating these stages, the steam was allowed to expand continuously: this method allowed for the creation of an engine that was much faster, lighter and smaller than the expansion engine.

In January 1894, the Marine Steam Turbine Company was formed in Newcastle-upon-Tyne to exploit the developments of Charles Algernon Parsons (1854-1931). In 1884, he had constructed a steam turbine to provide the motive power to drive a dynamo that generated electricity (*see* **The First Turbo-Generator**). In 1889, the partnership that had built the 1884 machine was dissolved. Under the terms of the partnership agreement, the patents relevant to the many improvements Parsons had introduced during the period 1884-89 became the property of his former partners. This arrangement meant that he was not allowed to design a turbine that incorporated parallel flow, that is, one that allowed steam to flow through the engine in the same direction as its shaft or axis.

To overcome this problem, Parsons resorted to radial flow, whereby the steam was introduced near the shaft and then made to flow radially outward before being conducted back inward to the shaft, where it could be used for another outward pass. It was this kind of steam turbine that was used to propel the world's first steam-turbine-driven vessel, *Turbinia*.

To estimate the power needed to drive *Turbinia*, Parsons conducted a series of resistance experiments with two models of the ship in early 1894. The predictions derived from the experiments were that the turbine would have to develop 820 horsepower at a boiler pressure of approximately 200 pounds per square inch in order to drive the vessel at thirty knots, using a single propeller shaft rotating at 1,600 revolutions per minute. Preliminary speed trials of the *Turbinia* were first conducted on November 14, 1894; the results were disappointing. A speed of just under twenty knots was achieved—ten knots below specifications.

In a further series of experiments, using a specially developed torsion dynamometer, which measured the power being created by the turbine and

The Turbinia *underway; Parsons is the tall figure just forward of the funnel*

transmitted by the propeller shaft, Parsons diagnosed the fault as inherent in the design of the propeller. Yet, although seven different designs of propeller were used in thirty-one trials, he could gain no improvement in overall performance.

After numerous additional experiments, Parsons found that the problem was that the drop in pressure over the back of the propeller blades caused the air that was dissolved in the water to form cavities. These cavities led to a drop in efficiency and a loss of thrust: this phenomenon was later called cavitation. Ultimately, cavitation was caused by the inability of the propeller to absorb the great amount of power being generated by the turbine. The reasons were the propeller's small diameter and its small surface area.

In order to overcome cavitation, Parsons decided to replace the single-turbine/single-propeller-shaft arrangement with three turbines, each driving its own propeller shaft. He made this decision toward the end of 1895, and it coincided with the recovery of his original patents. The radial turbine was then taken out; three parallel-flow reaction turbines were installed; and a newly designed propeller fitted. It was essentially this configuration that drove the *Turbinia* at 32.75 knots in February 1896 and that so impressed the Royal Navy when the vessel appeared at the Naval Review later that year.

Although the radial turbine was never used again after its trials on the *Turbinia*, and its design was superseded by the parallel (or axial) turbine, this particular engine was the seminal engine demonstrating not only the effectiveness of the steam turbine for marine propulsion, but also the need to understand the phenomenon of cavitation in the design of propellers.

—TOM WRIGHT

PANHARD ET LEVASSOR CAR (1895)

In Britain, Lanchester, Austin and others were experimenting with cars in the mid-1890's. However, from a European perspective, the successful invention, development and manufacture of automobiles was a Continental phenomenon. In Germany, Benz had demonstrated the first practical car in January 1886. The most successful gasoline engines were the high-speed engines built by Daimler, also in Germany, from 1884, and these engines were used in motor cars from 1886. But it was the French Panhard et Levassor car (using the Daimler engine) that transformed motoring into a serious proposition.

René Panhard (1841–1908) and Émile Levassor (1843–97) made woodworking machinery at their firm in Paris; they had acquired the French rights to the Daimler engine in 1889. They experimented first with cars in which the engine was in the center or at the rear of the vehicle. Then, in 1891, they developed and offered for sale a car in which a front-mounted, vertical, two-cylinder engine drove through a flywheel, clutch and three-speed gearbox to the rear axle. The driver and passengers sat between the engine and the rear axle. This arrangement, the *Système Panhard*, became the standard layout for the automobile for the next seventy years.

A car of this kind attracted worldwide publicity when it finished first in the first road race from Paris to Bordeaux and back on June 11–13, 1895. In a great feat of skill and endurance—given the machine he was driving—Levassor drove the car himself. He covered the distance of 732 miles in two days, forty-eight minutes, at an average speed of fifteen miles per hour; he arrived at the finish line six hours before the next competitor. Panhard cars were to dominate the great Continental road races until 1900.

Meanwhile in Britain, the embryonic motor-car industry was restricted by legislation that required motor vehicles to be preceded by a pedestrian who warned of their approach, effectively limiting their speed to four miles per hour. Evelyn Ellis (1843–1913) was one of a small number of pioneer British motorists who had the vision to see the importance of the motor car, and he campaigned to have the law amended. He chaired the inaugural meeting of the Automobile Club of Great Britain and Ireland (from 1907, the Royal Automobile Club), and he became its first vice-chairman. He was also a director of the Daimler Motor Co. Ltd. of Coventry.

On June 25, 1895, Ellis bought a Panhard of the kind that Levassor had driven in the Paris-Bordeaux-Paris race. It featured engine number 394 Type P2D of three and three quarter horsepower, with coachwork by Belvalette. It was the first car imported into Great Britain, and its first journey from Micheldever station to Ellis' home in Datchet on July 5th was vividly described by F.R. Simms in the *Saturday Review*:

"We set forth at exactly 9.26 a.m. and made good progress on the well-made old London coaching road…We were not quite without

Model of the 1894 Panhard et Levassor presented to Queen Elizabeth II in 1957 by the Royal Automobile Club

anxiety as to how the horses we might meet would behave towards their new rivals, but they took it very well, and out of 133 horses we passed on the road, only two little ponies did not seem to appreciate the innovation… It was a very pleasing sensation to go along the delightful roads towards Virginia Water at speeds varying from three to twenty miles per hour…There we took our lunch, and also fed our engine with a little oil…Going down the steep hill leading to Windsor, we passed through Datchet, and arriving right in front of the entrance hall of Mr. Ellis's house at Datchet at 5.40, thus completing our most enjoyable journey of fifty-six miles, the first ever made by a petroleum carriage in this country, in 5 hours 32 minutes, exclusive of stoppages."

Ellis claimed to have driven the Panhard for more than 2,000 miles over the next eighteen months, but he was never prosecuted for exceeding the speed limit or not being preceded by a pedestrian. It was one of the five cars at the first public exhibition of cars in Britain—at Tunbridge Wells on October 15, 1895. It also took part in the Emancipation Run from London to Brighton on November 14, 1896, which was held to celebrate the enactment of the Locomotives on Highways Act. This law freed motor cars from the restriction of being preceded by a pedestrian, and it raised the speed limit to twelve miles per hour. From that time, motoring in Britain was able to develop.

Ellis sold the Panhard pictured opposite for about £3 in 1908. It was acquired by Mr. A. Vaughan-Williams, a consulting engineer, who exhibited it in a display of historic cars at the Imperial International Exhibition at White City in London in 1909. Recognizing its importance, the Royal Automobile Club bought the car for £50 and presented it to the Science Museum the following year.

—PETER MANN

COMING SOUTH, PERTH STATION (1895)

Coming South, Perth Station is one of a pair of paintings by the Victorian artist George Earl (1824-1908). This painting and its companion *Going North, King's Cross Station* capture the spirit of the railway age at the height of its prosperity, before the advent of the motor car, when virtually everyone who traveled appreciable distances did so by train. Apart from William Powell Frith (1819-1909) in his seminal work *The Railway Station* (1862) and George Earl in these two works, no other Victorian painter was to represent in such panoramic detail the world of railway station interiors: their architecture, the trains, the wide range of passengers and their imagined stories, contemporary advertisements, and station staff.

George Earl is primarily known as an animal and sporting painter, who exhibited at the Royal Academy from 1856 to 1883, at the British Institution, and at the Society of British Artists. Although Earl was perhaps best known for his dog portraits, often in simple formats, he also attempted elaborate and sometimes dramatic compositions in which figures, animals, and landscapes were combined.

Earl originally explored the railway themes in two earlier works exhibited at the Royal Academy in 1876 and 1877. Later, reputedly at the request of Sir Andrew Barclay Walker of the Walker Brewery, he painted the two enlarged and amended versions of his original compositions. Both paintings came into the possession of the Walker Brewery Company, and until 1990 hung in one of the Walker pubs, The Vines, in Lime Street, Liverpool.

The companion paintings are fine illustrations of the social impact of the railway at the zenith of its power and influence in the late Victorian period.

The two paintings fit neatly between the romantic and sentimental canvasses of the 19th century and the impressionist works of the 20th century, both of which modes are already well represented in the National Railway Museum's picture collection, where *Going North, King's Cross Station* is housed.

In the painting, *Coming South, Perth Station*, dated 1895, the scene is Perth station, in Perth, Scotland. The fine train shed, designed by Sir William Tite in 1848 and lit by elegant gas lights, is a prominent feature. Although the station at Perth was enlarged in 1885-87, Earl shows it in its earlier condition with north and south trains leaving from the one hall. During conservation work on the painting, an examination of the brush strokes confirmed that Earl was responsible for depicting the architectural features of the station as well as the dogs and crowd scenes.

A train bound for London waits to depart. The first carriage is marked for London, Euston Station. A group of passengers, with its hunting dogs, game, bags and other equipment prepares to leave Perth following a successful shooting house-party. Among the luggage lie grouse, blackcock and some stags' antlers, while a porter heads toward the train with a rabbit and a wicker basket, doubtless containing salmon. Station porters are again in evidence, struggling with luggage on a trolley near the second carriage, the center portion of which would be the baggage compartment. A boy walks through the station selling *The Scotsman* newspaper, and the grooms and footmen control the dogs as best they can. On the left, an elderly woman in red, standing behind her husband who is wearing a tam-o'-shanter hat, appears upset at the departure of their daughter.

The station clock shows 3:50 p.m., which indicates that the train waiting to depart is the 4:04, arriving in London at 3:50 the following morning. The train would follow the West Coast route from Perth to London. The Highland Railway terminus is shown in the center of the painting; the Highland Railway train would leave at 4:30 p.m. for Inverness, arriving at 10:05 p.m.

Among the several gun dogs that Earl shows in the central group are pointers, which flank two English setters and a pair of Irish setters with gleaming tawny coats. Interestingly, there is a well-behaved border collie, profiled on the right, looking on, behind the cloaked woman: it is not a gun dog, and it accompanies the gentleman in the top hat who is perhaps present to witness the departure of a friend. The platform is littered with feathers, food, and an open book, possibly dropped by the woman in peach who is bending as though to retrieve it.

An interesting feature of the Perth painting is the prominence of the railway station signs. A notice at the end of the Highland Railway bay points people to the Caledonian Railway for through-trains to Aberdeen. The left-luggage room and the ladies first-class waiting room are clearly indicated. A little further down the platform, the ladies second-class waiting room sign can barely be seen. On the pillars alongside the platform are signs forbidding smoking at the station. This embargo is believed to have been an early instruction to passengers to ensure their safety; it was soon relaxed.

—CHRISTINE HEAP

THOMSON'S CATHODE-RAY TUBE (1897)

Matter is made of atoms. This concept is so familiar to us that it is now part of general knowledge, yet a century ago, some leading scientists doubted the very existence of the atom. The extraordinarily successful journey that led to our present understanding of atomic structure began in the rather dingy laboratory of Joseph John ("J.J.") Thomson (1856-1940). An enthusiastic and ambitious professor at Cambridge University's Cavendish Laboratory, Thomson used results he had obtained with his cathode-ray tube to prove in 1899 the existence of the first subatomic particle: the electron. In many ways, his discovery marked the birth of the modern electronic age.

Thomson was the apotheosis of the brilliant Cambridge scientist. He began his career by winning a scholarship to Trinity College, and he went on to be a Fellow, Professor, and Master of the College, winning along the way a Nobel Prize in Physics (1906), a knighthood (1908), and the presidency of the Royal Society (1915). He was Professor of Experimental Physics, although that may not have been the best job for him; he was very mathematically minded and was well known to be an inexpert experimenter, to the extent that his students became fearful when he approached their equipment (his wife would not let him do even minor jobs around the house). He was nonetheless a brilliant designer of scientific apparatus and equally good at interpreting the results of experiments. Both of these strengths were displayed in his conclusive contribution to the cathode-ray debate that was exercising European scientists in the 1890's.

In the late 1860's, it had been demonstrated that a column of gas at low pressure glows colorfully when electricity is passed through it from one metal plate to another. If the pressure is sufficiently low, the individual tracks of so called "cathode rays" can be seen between the plates (*see* **Crookes' Radiometer**). Of what did these rays consist? If that question had been put to a European scientist in

J.J. Thomson with a cathode-ray tube, 1897
(The Cavendish Laboratory, University of Cambridge)

the mid-1890's, his answer would have depended to some extent on which side of the Rhine the scientist lived. German scientists held that the rays were mysterious waves in the ether that pervaded the whole of space, whereas French and British scientists believed that the rays were comprised of individual particles.

In the race to rule out the "wrong" theory of cathode rays, Thomson made a key contribution in the spring of 1897. With the help of his assistant, Everett, he constructed special cathode-ray tubes that enabled him to study how the paths of the rays were affected by electric and magnetic fields.

In his Royal Institution lecture of April 30, 1897, Thomson boldly—some might say rashly—suggested that cathode rays were corpuscles even smaller than hydrogen atoms. At least one member of his audience thought he had said this in jest. But he certainly did not, and two years later, he went further and suggested that cathode rays were indeed particles, each with a definite charge and mass (his values for both turned out to be quite accurate). He had written his name into every book on atomic science that would henceforth be written: he was the discoverer of the electron.

This may have been a great event for the scientific cognoscenti, but the question then was: would it make any difference to the lives of the public at large? Some scientists evidently hoped not; for some years after Thomson's breakthrough, his colleagues at the Cavendish, at their annual dinner, toasted: "The Electron: may it never be of use to anybody." This pompous disdain has not been justified—electrons are now very much a part of modern life. They illuminate the screens of every television and personal computer, and they are responsible for the operation of every electrical device. In laboratories, electron microscopes allow mapping of the shapes of viruses and the surfaces of atoms (both far too small for ordinary optical microscopes); the electronics industry is always striving to exploit new and more ingenious ways of controlling electron flow; and radiologists routinely use electrons from radioactive decays to diagnose and treat cancers.

Yet, although the aspirations of the Cavendish purists have been thwarted by the technicians, others have advanced knowledge of the subatomic world in ways that are beyond Thomson's dreams. The electron is now known to be just one of several fundamental particles, each envisioned to exist at an unimaginably small point, with neither shape nor size. Scientists now study these basic building blocks of matter with huge particle accelerators that are each, in a sense, descendants of Thomson's little cathode-ray tube. Today's particle physicists normally work in multinational teams, and can take years to design and implement experiments that consume large chunks of national research budgets. The days in which a lone scientist could make a revolutionary discovery of a new particle seem, sadly, to have passed.

—Graham Farmelo

It has been said that Mr. Marconi has done nothing new. He has not discovered any new rays; his transmitter is comparatively old; his receiver is based on Branly's coherer. Columbus did not invent the egg, but he showed how to make it stand on its end…
—William Preece, Friday Evening Discourse at the Royal Institution, June 1897

In 1896, Annie Marconi, the well-heeled daughter of the Jameson Irish Whiskey family, brought her son Guglielmo Marconi (1874-1937) from Italy to England to seek a sponsor for his new system of wireless telegraphy, the Italian Ministry of Posts and Telegraphs having shown little interest in it. His first success was with the prickly and conservative Engineer-in-Chief of the Post Office, William Preece, the man who had once dismissed Bell's telephone (*see* **Bell's Osborne Telephone**) with the famous remark, "I have one in my office, but more for show, as I do not use it because I do not want it."

Preece became Marconi's champion, arranging and publicizing tests of wireless; he even renounced his own rival system. Marconi clearly had something that Bell, and several other pioneers to whom Preece offered less than wholehearted support, did not. Whatever it was, this relatively untutored youth progressed from experimenting in an attic in Bologna to producing an unglamorous but tuned (and in that sense modern) transmitter in just five years.

Heinrich Hertz (1857-94) had confirmed the existence of electromagnetic waves in 1888, generating them with the help of sparks. A year later, Oliver Lodge (1851-1940) showed how an electric circuit could be made to vibrate like a plucked string and make a nearby circuit vibrate in sympathy. In 1894, when he was twenty, Marconi began experimenting with Hertz's waves, with the definite intention of using them as tools of communication.

Five years on, with wireless threatening to degenerate into Babel because of the problem of interference, Marconi was exploiting Lodge's ideas to produce a solution. The qualities that attracted Preece to Marconi, perhaps because he shared them, were evident throughout. Weak on theory, but with a clear vision of his goal, Marconi groped his way to the realization of what academics had merely shown to be possible. Prototypes were his indispensable guides.

The tuned transmitter emerged from Marconi's work on detecting rather than on creating electromagnetic waves. From the beginning, he had used as part of his receiver the coherer (used for detecting radio waves), a well-known but capricious device that he converted into a reliable component. But, at the end connected to the receiver, his aerials produced large currents and small voltages. The coherer needed the opposite effect.

He found a solution by connecting the coherer through a special electrical transformer, which his staff called the "jigger." It was the outcome of a program of trial and error that began in 1897. By 1898, the jigger had increased by a factor of nine the distance over which Marconi could communicate—to sixty miles. Such sensitivity had its disadvantages. There could be many stations in a sixty-mile radius,

Marconi's first tuned transmitter, 1897

and if more than two started transmitting at the same time, nobody would receive anything intelligible.

Someone in a crowd can concentrate on one speaker at a time, ignoring other conversations. The problem was how to give a radio receiver similar competence. Marconi was not alone in realizing that, with a transmitter producing waves of well-defined frequency, the electrical resonance that Lodge had demonstrated could make a receiver respond strongly to that transmitter and to no other. The difficulty was in finding a means of producing electromagnetic waves at a desired frequency. Solving this problem was the final step toward tuning, as well as toward modern radio communication. The jigger transformer, originally designed to improve reception, provided the means.

Like Hertz, Marconi used an electrical spark as a means of releasing radio waves. With the spark gap connected to the aerial, as in the early Marconi transmitters, radio waves were produced in bursts too brief for any definite frequency to be discernible. By connecting the gap through a transformer—the few turns of wire wrapped round the frame of his prototype—each spark could produce a train of waves long enough to excite resonance. If all the circuits involved were tuned to what Lodge called "syntony," transmitter and receiver could be alone in a crowd.

William Preece, who may have been wrong about some other inventions, was right about Marconi, the gifted magpie. In fulfilling his vision, Marconi used whatever and whomever he could, but he did so in a way that an artist or scientist must use his materials; that is, without prejudice. He did not stumble upon the tuned transmitter, but he did not arrive at it quite rationally either. It was first revealed in his Patent No. 7777 of 1900. Numerologists may make of that what they will, but without doubt this humble transmitter is a seminal fragment of the engineer's art.

—Roger Bridgman

POULSEN'S TELEGRAPHONE (c. 1903)

In the days before the digital age, those who regarded the telephone answering machine as the quintessential curse of modern existence might have been surprised to learn that it was invented by a relatively obscure Dane as long ago as 1898. In the process, Valdemar Poulsen (1869-1942) also invented magnetic recording, the foundation of what became a huge industry encompassing tape, disks and indeed answering machines themselves. He did not conceive the idea, or perfect it, but he was undoubtedly the first person to create a working device.

When Poulsen was born in Copenhagen, few had ever heard speech from the mouth of a machine. By the time he died in 1942, magnetic recording, while not yet commonplace, was steadily becoming so. In the late 19th century, however, few could have attempted such an invention. Poulsen's only option then was to translate every subtlety of the sound wave into a corresponding fluctuation of the magnetic record, and to do that properly he needed amplifiers of a kind not then available. Digital recording, the idea that sound waves could be reduced to numbers, unsubtle and easy to record, was not an idea that was put forth seriously until just five years before his death.

At the time Poulsen revealed the machine that he was eventually to call the telegraphone, the culture of the industrialized world was still dominated by the written word. Edison at first thought of his own recording machine, the phonograph, as a device that would allow telephone calls to be handled by a central office, as telegram messages were handled by a post office. Poulsen's attitude was more democratic. In his first British patent, he describes machines that would sit in a home waiting for an absent owner, announcing the time of the owner's intended return to callers and, if required, taking messages.

Between 1898 and 1915, Poulsen produced many modifications of his original idea. His first machine was clearly a magnetic version of the phonograph, using a cylinder wrapped with steel wire that became magnetized in a pattern similar to the

Valdemar Poulsen never earned a degree but was awarded two doctorates for his pioneering work in speech communication (International Telecommunications Union)

sound waves. He later used reels of wire for recording incoming calls, and he also considered steel tape, disks, and even magnetically coated paper—but he struggled with the problems of handling these media.

As well as lacking amplification, his machines were complicated and probably unreliable. They achieved some success, however, for office dictation. Indeed, their quality of reproduction gained Poulsen the Grand Prix of the Paris Exposition of 1900, though it must be said that the only contemporary objects of comparison were the horribly scratchy phonograph and gramophone.

Although no records of the manufacture or sale of the telegraphone pictured opposite have survived, we know that it was used in one of the Royal Dockyards, and its design indicates a date of 1903 or 1904. The machine used steel wire, driven at about fifty times the speed of tape that was to be

used decades later. It could have been used as either an answering or a dictating machine. It would have been supplied with a pair of Bell-type telephones, and both had to be pressed to the ears to hear speech at intelligible volume. Still, users were impressed by the absence of background noise, making possible the recording of whispers and sighs, something the rival phonograph could not match.

Magnetic recording did not really become useful until three fundamental problems had been solved—and Poulsen never really tackled any of them. First, a large amount of amplification was needed. The power from the playback head that was used in a typical cassette-tape system was increased 100 billion times on its way to the speakers. There also had to be a medium that would retain fine magnetic detail, so that the recorded wavelength of sounds could be reduced and, with that reduction, the absurdly extravagant wire utilization in the telegraphone could also have been reduced.

Finally, to have achieved an undistorted recording, high-frequency bias was required—in order to compensate for the fact that magnetic media did not react in the same way to both large and small signals. Poulsen failed to solve this problem completely, and a solution was not found until 1927. Paradoxically, because they are digital, modern recordings do not require this solution.

Ultimately, the telegraphone was a failure; its purpose was too far in advance of the technology needs of its day. A truly satisfactory answering machine was not available until the 1950's (like the telegraphone, it recorded on wire), and it was thirty years after that before many people became more adept at speaking to answering machines and voice message systems than they were to actual human beings. Ever the democrat, Poulsen gave up recording and instead developed the first radio transmitter capable of handling speech. This pre-electronic device enjoyed only brief acclaim, but the technology that replaced it was responsible for turning his other little failure into a huge success.

—ROGER BRIDGMAN

Distress and disappointment hardly seem appropriate rewards for the holder of one of the most significant patents in the history of electronics. Yet, they were the misfortunes that befell John Ambrose Fleming (1849–1945) after registering his patent for "Improvements in Instruments for Detecting and Measuring Alternating Electric Currents" in 1904. This title, and the ensuing unpleasantness, obscures the fact that Fleming's thermionic valve (or vacuum tube, or electron tube) was the first practical thermionic device for detecting radio signals. Thermionic devices are those in which the source of current is a heated negative electrode (the cathode), and the current flows through a vacuum to the positive electrode (the anode) via other electrodes that can be used to control the current.

Fleming's valve was the first of a line of devices that were to be the mainstay of electronics for the first half of the 20th century and well into the solid-state era, when transistors and integrated circuits (semiconductors) inspired the miniaturization that has become an expectation. Today, thermionic devices are not nearly as prevalent, although there are still some cathode-ray tubes in television and computer screens (see **Crookes' Radiometer**) and high-power valves in transmitters. As well, many audiophiles actually prefer the "tube sound" to solid-state amplifiers, due to the texture of the sound as the tube-based amplifier is pushed to maximum power.

John Ambrose Fleming became Professor of Electrical Engineering at University College, London, in 1885. Although an academic, he was the originator of what might be called applied electronics. In 1882, Fleming had become "electrician" to the Edison Electric Light Company, and he was soon absorbed with the task of improving carbon-filament light bulbs, which had a tendency to darken. Thomas Edison (1847–1931) himself had already tried, unsuccessfully, to prolong bulb life by placing an extra electrode alongside the filament: he noticed that, with an extra electrode connected to the positive end of the filament, a small but measurable electric current flowed between them. This

"Edison effect" caused widespread curiosity, and Edison patented his modified lamp as a novel "electrical indicator."

Fleming commenced more serious research into the Edison effect in 1889. He arranged for experimental lamps to be made at the Edison and Swan (later Ediswan) lamp works. The results obtained with them, reported in 1890, led to further work in 1895 and 1896. By then, Fleming had investigated electrical conduction in a vacuum between glowing filaments and a second electrode in a variety of shapes and sizes of tubes.

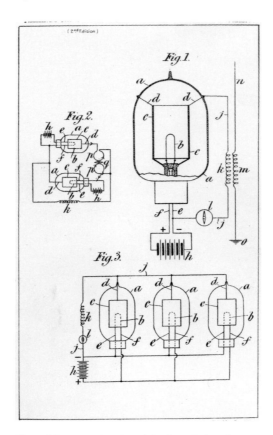

Part of the patent drawing of the first wireless detector to use a thermionic valve. Opposite: an experimental lamp which Fleming used in 1889. It was to one of these that he turned fifteen years later for his wireless detector experiments

His painstaking account of this work is now regarded as a classic. But in a world that was more excited by Röntgen's revelations on x-rays and J.J. Thomson's identification of the electron (see **Thomson's Cathode-Ray Tube**), Fleming's work was often overshadowed.

But his moment was yet to come. Wireless telegraphy was expanding rapidly (see **Marconi's First Tuned Transmitter**). Fleming became a technical adviser to the Marconi Company in 1899, and he helped design the transmitter at Poldhu, Cornwall, for the 1901 transatlantic transmissions. There was a significant problem that had yet to be solved: receiving weak wireless signals, especially over long distances. Fleming, inspired in 1904 by "a sudden very happy thought," turned to one of his earliest 1889 experimental valves for a solution. He linked it up to a simple circuit containing a battery, a meter, and suitable coupling to an aerial. He found that the system worked very well.

A patent was granted in 1905, but any benefits it might have brought Fleming became the property of the Marconi Company. "Cat's whisker" detectors appeared a year later: they were less cumbersome and equally efficient. And in 1907, the American Lee De Forest patented a detecting circuit using a valve with a third electrode. Fleming was well aware of De Forest's activities, and believed that De Forest had copied his ideas. However, De Forest persisted, and a lifelong enmity between the pair ensued. It was not until 1912, however, that the full potential of the extra wire in De Forest's valve was realized, when it was incorporated into circuits that could generate and amplify speech and radio signals—a development that took communications firmly into the electronic age. But without Fleming's persistent work, this transition would have been delayed—or credited elsewhere.

The final, sad twist to the story is that the enmity between the Fleming and De Forest factions in the United States spilled over into patent litigation, and in 1943, the U.S. Supreme Court ruled that Fleming's American patent was invalid.

—Eryl Davies

MASS PRODUCTION AND THE MODEL T FORD (1908)

The Model T Ford was the sturdy, well-priced car that motorized America. But it was much more, for it represented the claim that Henry Ford (1863-1947) had to have discovered the technique of mass production and even to have rewritten the laws of economics through "the Ford prosperity recipe... high wages, low prices, and mass production."

The broader history of mass production is more intriguing than that of just the automobile, for mass production had deep historical roots and a long, vexing development. In 1790, a French inventor, Honoré Blanc, exhibited bins of musket parts and assembled them at random into functioning weapons, to show that they were identical and perfectly interchangeable. A systematically organized state-run workshop, he claimed, would "sweep away the ancient regime" of hand-skills and reticent craft gunsmiths.

The new American republic was fertile ground for French techniques. Stimulated by reports from Thomas Jefferson in Paris, idealized "interchangeable" firearm manufacture became an objective for the new American army.

These early achievements were orchestrated to win government support. An example is Eli Whitney (1765-1825), who also submitted sets of firearm parts for inspectors to mix up and reassemble as they pleased. This demonstration argued for the exactness of his production, but, as David Hounshell has shown (*From the American System to Mass Production 1800–1932*, 1984), Whitney's works involved only the simplest, least expensive equipment: extensive selection and trial assembly lay behind his interchangeable demonstration sets.

Even Colonel Samuel Colt (1814–62), who became identified with exact repeatability in production, was rather inexact before an 1854 British parliamentary committee on small arms, stating that interchanging parts between weapons "would not cost you very much." His former production supervisor went even further, saying that "I have heard of it (perfect interchangeability), but I defy a man to show me a case."

The reasons for this slow development of repeatability are complex. Tools wear and change as they cut, stamp, or forge parts, so that there will be an inevitable dimensional "creep" as well as a random statistical variation. Moreover, the unambiguous definition of the geometry of a three-dimensional part was not a trivial problem; neither was the development of micrometer measurement. Ideal absolute sizes were not achievable. The philosophy of gauging, and standards for what size ranges ("limits" or "tolerances") were acceptable, took time and many experiments.

Initially, true interchangeability was expensive. In arms, consistency couldn't be achieved by a high rate of scrapping of unusable parts. Mass production for the marketplace required low waste and progress in development of rigidity, repeatability and consistency in machine tools.

Ford cars, in fact, could not have been made cheaply at any earlier point in history. And though the "American System" (as evolved by New England armory practice) played a part, there was much more involved in the development of mass production. "Moving the work to the men" was one ingredient, and, in a throwaway remark, Ford suggested that the "disassembly" lines of the Chicago meatpackers were an inspiration, but conveyors had already been a part of the industrial milieu in the United States.

Another ingredient was the ruthless subdivision of labor: "the man who places a part does not fasten it...the man who puts on a nut does not fasten it." But Frederick Winslow Taylor (1856-1915), the American mechanical engineer and efficiency expert, and his theories regarding "scientific management" may have been less influential than many argue. The transformation for the Ford plant was far more subtle and organic than Taylor could have arranged. While Ford concentrated on detailed improvements to "work steps," his production men used a deep understanding of engineering possibilities, which Taylor did not possess. They designed parts to cut out process steps or to reduce criticality in particular dimensions, experimented with new materials and faster metal-forming techniques, and organized the sequencing of operations and the precise location of machine tools in the plant.

Charles Sorensen, one of Ford's experts, was explicit in crediting this team: "Henry Ford had no ideas on mass production. He wanted to build a lot of autos. He was determined but, like everyone else at that time, he didn't know how. In later years he was glorified as the originator of the mass-production idea. Far from it; he just grew into it like the rest of us. The essential tools, and the final assembly line with its many integrated feeders, resulted from an organization that was continually experimenting and improvising..."

The production line ultimately realized the vision of the Enlightenment for systematic rational manufacture; intriguingly, Ford and his methods impressed both Hitler and Lenin.

As Simon Schaffer has shown ("Enlightened Automata" in *The Sciences in Enlightened Europe*, 1999), one of the lesser-known ambitions of Enlightenment thinking was "to turn men into machines"—to extract the knowledge of the secretive craftsman and put the workplace under the direction of "a philosopher." Adam Ferguson, the Scottish historian, wrote that "mechanical arts... succeed best under a total suppression of sentiment and reason...Manufacturers prosper where...the workshop may...be considered as an engine, the parts of which are men." That is probably a perfect description of Ford's Highland Park, Michigan factory.

—ANDREW NAHUM

HABER'S SYNTHETIC AMMONIA (1909)

Both science and good engineering are important aspects of technological innovation. Nowhere is this lesson better demonstrated than in the story of the early 20th-century invention of synthetic ammonia, which became the basis of modern fertilizers. The two key relics (both pictured) of this German breakthrough are a tiny glass tube holding a few cubic centimeters of liquid and a thirteenmeter, sixty-ton steel pressure mantle. The sample in the tube was produced when the laboratory process of creating synthetic ammonia was first demonstrated on July 2, 1909. The giant pressure mantle is a relic of the industrial realization of the physical chemist's dream, and it was part of a plant that operated from 1916 to 1982.

Ammonia had been compounded into fertilizers since the mid-19th century, enriching the land, enabling year-in and year-out food production for the world's growing population. But by the 1890's, it appeared that the main source, guano from Chile, would last no more than twenty years; the prospect for its absence was famine. Among the scientists replying to this challenge, the physical chemist Fritz Haber (1868-1934), of the University of Karlsruhe, calculated that it would be possible to obtain ammonia by simply "burning" water, since ammonia's two ingredients, nitrogen and hydrogen, are found in both air and water. To convert his vision (even into just a laboratory demonstration), Haber needed a new understanding of physical chemistry, high pressures, catalysts made from newly-available osmium, and a well-designed apparatus.

The apparatus was provided by one of Haber's student-assistants, Robert Le Rossignol, who was from the U.K. Channel Islands. He designed a system that successfully produced synthetic ammonia. On July 2, 1909, the process was demonstrated to Alwin Mittasch of Badische Anilin- & Soda-Fabrik (better known today as BASF). Although only eighty cubic centimeters were produced that day, the principle had been proven, and the company thereafter supported the commercial development of the process.

Transforming an interesting laboratory demonstration into a commercial process destined to be the forerunner of all high-pressure chemical technology was as formidable a task as the practical demonstration of the original theory. BASF believed that developing the process was a major commercial opportunity in agriculture, therefore they were willing to proceed—but another commercial opportunity presented itself once World War I started in August 1914. Ammonia was the most practical source of the nitric acid needed to make TNT, the favored general explosive. A 1914 British naval victory, the Battle of the Falkland Islands, meant that Germany was cut off from Chilean guano, from which it had derived its primary supply of nitrates used to make ammonia.

A pressure mantle being installed in the Oppau plant, 1920 (BASF)

The scale of Germany's daily ammonia requirements continued to grow, first to one ton, then to twenty, then, by 1916, to 150 tons a day. The challenge was to build the immense pressure reactors, each capable of producing twenty tons of ammonia a day. The combination of Haber's scientific achievement and the engineering project headed by Carl Bosch (1874-1940) of BASF had, by 1918, produced two full-scale plants: at Oppau, near Ludwigshafen am Rhein, and at Leuna, near Leipzig. Each could produce 50,000 tons annually, which met half of Germany's ammonia needs.

Wartime shortages meant that a new catalyst had to be found, and German scientists developed a new iron-based material. Experimenters also discovered that the hot hydrogen used to make ammonia reacted with the carbon in the steel of the reaction vessels at the enormous pressures needed, 200 times atmospheric pressure. A special lining of low-pressure carbon steel was necessary, and to ensure that this lining was not itself put under undue pressure, the still cool but pressurized gases were circulated behind it.

The pressure mantles from the 1916 Oppau plant were made of two immense six-and-a-half-meter forgings bolted end to end. The thickness of the steels walls—twelve and a half centimeters—rose to thirty centimeters at the flanges. Along with the vessel, there was an electric heater that brought temperatures up to the 200 degrees centigrade required.

Haber, in addition to his work on synthetic ammonia, also won notoriety for his work on poison gases which were used during World War I. He won the 1918 Nobel Prize in Chemistry for the synthesis of ammonia from its elements. In 1931, Carl Bosch also won the Nobel Prize in Chemistry for his contributions to the invention and development of chemical high-pressure methods. The process that the two men developed is still in use today.

—ROBERT BUD

THE MUNITIONS GIRLS (1918)

"The people that will be able to keep its forges going, will perforce be the victor, for it alone will have arms," the French commentator Messance said about war in 1788. In August 1914, the beginning of World War I, both the likely scale and duration of the war were underestimated. Winston Churchill's description, "A Steel War," in his Ministry of Munitions *Circular to Steelworks* of September 1917, captured both the reality of the situation and the truth of Messance's prediction: during the Great War, the iron and steel industry of South Yorkshire was turned over almost totally to the satisfaction of military needs.

The painting *The Munitions Girls* by Stanhope Forbes, commissioned in 1918, depicts the manufacture of four-and-a-half-inch shells in the Kilnhurst Steel Works of John Baker & Co. (Rotherham) Ltd., near Sheffield, England. It is a record of an industrial process: red-hot steel billets are reheated in an intermediate furnace and, still glowing, are moved by hand to a press for forging.

The giant and powerful press in the center of the painting might seem ill-fitted for its purpose. The artist has recorded machinery originally designed for the production of forged tires and axles for railway wagons; it has been converted for shell making. In 1915, as with many other steel makers, John Baker & Co. urgently had to retool for armament manufacture and, in particular, to meet the overwhelming demand for shells for the Western Front. Such was the demand that, having refitted its own plant, John Baker & Co. resorted to buying cattle-cake presses and altering them for shell manufacture.

The dominant theme of the picture, however, is the workforce: women—the "munitions girls" of the title. The painting was commissioned by George Baker, managing director of the firm, primarily as "…a memento for our women workers, and each of them received a framed copy of it." It is a graphic record of the wider employment of women in skilled trades. The call to the front had stripped the iron and steel industry of skilled male workers. Women left their traditional tasks, such as

file cutting, and moved into the machine shops and foundries, operating equipment formerly handled only by men. Even the hot and heavy work of the foundry was undertaken by women, though Forbes' painting seems to indicate that they were required in large numbers to compensate for the muscle strength of their male counterparts. During the four years of war, the number of women in the Sheffield steel industry increased six-fold. The 42,000 women in the country's steel industry by the end of the war represented eleven percent of the total workforce—some 36,000 more than before the war. They made a vital contribution to alleviating the labor shortage.

Both munitions manufacture and women's work were subjects covered by official war-art commissions from 1916, which included artists of industrial activity, such as C.R.W. Nevinson and

Recognition at Burlington House; The Munitions Girls *was exhibited at the Royal Academy Summer Exhibition in 1919. It gave rise to this cartoon in which a couple of society girls recognize as studio models the two figures whom Forbes had placed in the right foreground of the painting*

Muirhead Bone, but women's work efforts received little official artistic attention during the war compared to that afforded to men. In 1918, the first official British woman war artist, Victoria Monkhouse, was commissioned to record women in their new working roles—but to depict them as clean and picturesque. Iron and steel companies in Sheffield and Rotherham independently commissioned artists to record their own war efforts, building upon a longstanding tradition of artistic patronage for depictions of foundry and manufacturing scenes. A rolling mill, and women making files are among scenes by E.F. Skinner from a set of paintings commissioned in 1917 by Cammell Laird in order to aid the Red Cross. Yet, Skinner's *For King and Country* of 1919, now in the Imperial War Museum, London, is primarily an apotheosis of the heroic woman shell-worker rather than an acknowledgement of women's contributions to the industry.

Stanhope Alexander Forbes (1857-1947) was not an official war artist. After study in London, work in Paris, and studies in Brittany, he settled in Newlyn, Cornwall in 1884, founding the influential Newlyn School of Art in 1899. An exhibitor at the Royal Academy in London from 1878, he was elected an Academician in 1910. *The Munitions Girls*, exhibited at the Academy during the summer of 1919, contrasts markedly in subject and tonality with the landscapes, coastal and town scenes, and figure studies for which Forbes is primarily known, but it retains the narrative content and social realism that were his hallmarks.

Forbes painted several industrial subjects, among which were *Forging the Anchor* (exhibited in 1892) and paintings of the Sandberg sorbitic steel process. These works may have encouraged George Baker to commission *The Munitions Girls* in 1918. Writing to Mrs. Baker in 1940, Forbes recalled: "It was indeed an unforgettable sight to see those fine women carrying on their work so splendidly, and the opportunity which Mr. Baker gave me to record this wonderful service is one for which I can never be sufficiently grateful."

—WENDY SHERIDAN

Francis William Aston (1877–1945) was presented with the Nobel Prize in Chemistry in December 1922 for his work in establishing the isotopes of the non-radioactive elements. Isotopes are atoms that have the same chemical properties, but different atomic mass. For example: carbon can exist in its common form C^{12} or the less common C^{14}. At the award ceremony, it was acknowledged that radioactive elements had readily shown the existence of isotopes; however, proving that non-radioactive elements also possessed isotopes had been far from easy.

J.J. Thomson (*see* **Thomson's Cathode-Ray Tube**), himself a Nobel Laureate, had achieved inconclusive experimental results to this end, but he did so with the quartz microbalance and mass spectrograph. Aston, who was Thomson's assistant, identified two isotopes of the element neon and went on to employ his original mass spectrograph to find the different isotopes of about fifty elements.

Aston was the second son of a Birmingham metal merchant and farmer, and in 1893, he began to study science under Tilden, Frankland and Poynting at what later became Birmingham University. Although he continued with his isotope research for the next twenty years, it was his appointment by Sir J.J. Thomson at the Cavendish Laboratory in Cambridge that led to the discoveries for which Aston is now best remembered.

Aston was assigned to improve Thomson's apparatus in which a beam of positively charged particles (positive rays) were deflected by a combination of electric and magnetic fields into sharp visible curves, each representing an individual particle's charge-to-mass ratio. He thought that this apparatus gave rigorous proof that all the individual molecules of any given substance had the same mass. This Daltonian belief was rudely shattered in 1912 when Thomson obtained two curves for neon corresponding to masses 20 and 22. Two explanations suggested themselves: if neon had a true atomic weight of 20 (instead of the previously agreed figure of 20.20), then either mass 22 was an unknown compound of neon, or a new element: meta-neon.

Aston was assigned to investigate the latter possibility, and he tried to separate the meta-neon by a variety of techniques. To see how well he was succeeding in separating neon and the new mysterious substance, he devised the miniature quartz microbalance. The arm of the balance was level only when the case surrounding the balance was filled with gas of known density. Aston then filled it with the unknown gas, and altered its pressure, thereby varying its density until the tiny beam balanced again. Comparing the pressures allowed him

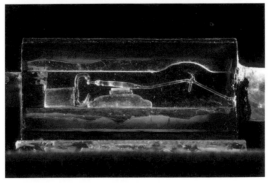

Top: Aston with his third mass spectrograph (The Cavendish Laboratory, University of Cambridge). Above: Aston's quartz microbalance

to calculate the atomic weight of the unknown gas. The results indicated that the mysterious substance was an element with the same properties as neon— but with a different atomic weight.

The coming of war delayed further experiments, but, on returning to the Cavendish Laboratory in 1919, Aston attacked the problem from a different direction, building a positive ray, or mass spectrograph. The key advance over Thomson's apparatus was Aston's arrangement of the electric and magnetic deflecting fields so as to bring rays of uniform charge-to-mass ratio to sharp focus on a photographic plate. Aston devised several methods for calibrating his instrument, and in the case of neon he obtained mass lines on his photographic plate at 20 and 22. The intensity of the lines showed that the two particles occurred in the ratio of 10:1—consistent with an average mass of 20.20, the known atomic weight of neon. Neon was thus proven to be isotopic, and in the short time before Aston was presented with his Nobel Prize, he had demonstrated the existence of isotopes in some thirty other gaseous elements.

Aston's work provided important insights into the structure of the atom and in the way that different elements are related to each other; for these reasons, he received the ultimate accolade of the scientific community.

With later, more accurate mass spectrographs, Aston obtained valuable information on the stability and abundance of particular isotopes. As a result of his work on the structure of the atom, he foresaw the dangers should the incredible energy of the nucleus ever be uncontrollably released, and he lived long enough to see his forebodings given awesome substance. Meanwhile, his principal experimental tool, the mass spectrograph, has been refined almost beyond recognition, adapted and adopted as an analytical instrument of prime importance in innumerable areas of chemical and biological research and industrial practice.

—DEREK ROBINSON

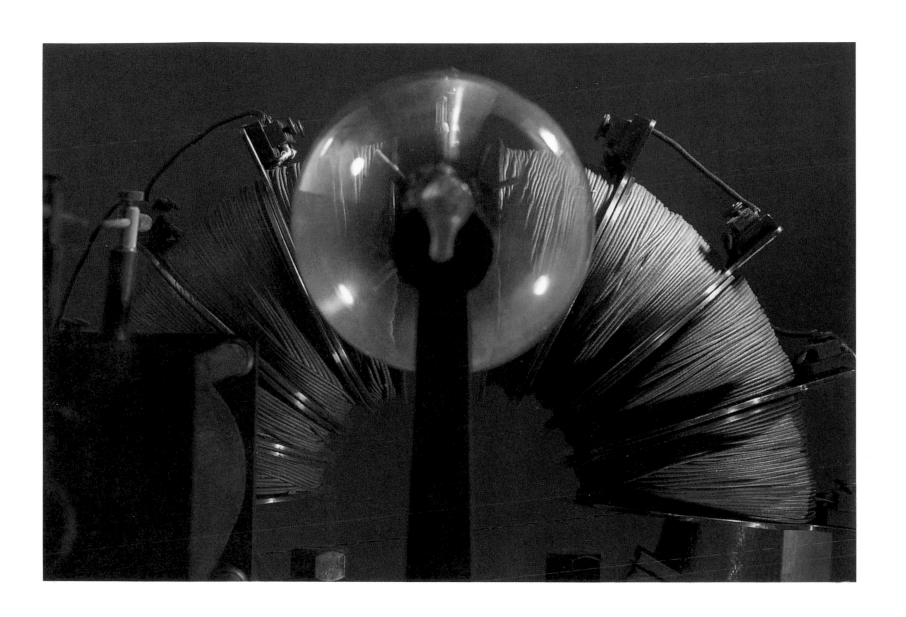

ALCOCK AND BROWN'S VICKERS VIMY (1919)

In 1913, London's *Daily Mail* newspaper offered a prize of £10,000 to the first aviator to fly across the North Atlantic. The outbreak of war in 1914 prevented any attempts to win the prize, but the offer was revived in November 1918. The Vickers company decided to enter the contest in February 1919, using a twin-engine Vimy aircraft flown by Captain John Alcock (1892-1919) and Lieutenant Arthur Whitten Brown (1886-1948), a pilot and navigator respectively.

The Vimy was designed as a heavy bomber to attack enemy targets in Germany, in response to the German Gotha bomber raids on London in 1917. Although the prototype Vimy was flown in November 1917, a shortage of engines prevented deliveries of production aircraft to the Royal Air Force until February 1919, after the Armistice.

The Vimy was a conventional design for its time, built mainly of wood with fabric covering, but careful attention to detail design and weight-saving produced an efficient structure capable of carrying the large fuel load needed for an estimated flight time of some twenty hours. The two Rolls-Royce Eagle VIII engines were probably the most reliable in service at the time; most contemporary engines needed a complete overhaul after about thirty hours of running—but the Eagle's standard time was a hundred hours. That was perhaps the most important technical factor in the Vimy's successful Atlantic crossing.

The aircraft selected for the flight was modified during construction at Vickers' Brooklands factory, mainly to increase the fuel capacity from 516 gallons to 865, providing for a nominal range of 2,440 miles. All military equipment was removed, and the pilot's cockpit was altered to accommodate the pilot and navigator side-by-side on a narrow, flat wooden bench, softened by a thin cushion.

After a single test flight at Brooklands, the Vimy was dismantled and shipped to St. John's in Newfoundland, where it arrived on May 16, 1919. The assembly of the aircraft in the open by a team of ten workers from Vickers and a single Rolls-Royce engineer, using improvised scaffolding and lifting gear, was no small feat. Finding a suitable field from which to fly was also a problem; eventually, by demolishing a stone wall and removing several trees, they created a sloping run of 500 yards.

The aircraft was ready to leave on June 14th. Alcock and Brown arrived at the field just before dawn, but they then had to wait for ten hours for the wind to subside. They were finally airborne at 1:41 p.m. (16:12 Greenwich Mean Time), climbing slowly to the west before turning back onto their course for Ireland and crossing the coast at 16:28 GMT. For eight hours, the flight proceeded between cloud layers above and below, but eventually Brown was able to get sextant sights on two stars and fix their position, which was well ahead of his estimate; they were flying with a tail wind of forty miles per hour.

Three hours later, Alcock lost control while flying through dense clouds at about 4,000 feet, and the Vimy fell into a spiral dive, emerging from the clouds in a steep bank. Luckily, Alcock was able to regain control just above the sea and resume course. At about 05:30 GMT, heavy snow obscured the fuel-flow sight gauge, and Brown had to climb up from his seat and kneel on the fuselage to clear the glass and confirm that the fuel pumps were still operating; he had to repeat this excursion, from the relative warmth of the cockpit into the blizzard, several times.

At 07:20 GMT, Brown obtained a sun sight, fixing their position near the Irish coast. At 08:25 GMT, they crossed the Irish coast near the Marconi radio station at Clifden, about twenty-five miles north of their intended landing spot. However, after sixteen hours in the air, they were anxious to finish their trip, and Alcock decided to land in a large green field not far from the radio station. Unfortunately, the field was actually a bog—but the Vimy touched down at 08:40 GMT, coming to an undignified halt with its nose in the mud.

Alcock and Brown returned to an enormously enthusiastic public welcome, and they were knighted by King George V. The Vimy was dismantled and returned to Brooklands, where it was rebuilt using new components to replace broken parts. It was not flown again, but was presented by Vickers Ltd. and Rolls-Royce Ltd. to the Science Museum in December 1919. Over the next few years, other Vimys made several record-breaking long-distance flights, including England to Australia and England to South Africa. The design also formed the basis for future commercial airliners.

—JOHN BAGLEY

AUSTIN SEVEN PROTOTYPE CAR (1922)

Many attempts to produce a small, reliable and economical car had been made since 1910. But it was not until the appearance of the Austin Seven, in July 1922, that this kind of car became practical. Sir Herbert Austin (1866-1941) was clear that the car should be aimed at the person who could afford only a motorcycle and sidecar but who aspired to be a car owner. He said that this new car would "knock the motorcycle and sidecar into a cocked hat and far surpass it in comfort and passenger-carrying capacity…I cannot imagine anyone riding a sidecar if he could afford a car."

Adding the sidecar to the motorcycle in 1903 had brought family motoring within the reach of those who could not afford the much more expensive motorcar. Cycle-cars, based on motorcycle engines and parts, had begun to appear around 1910. In spite of their crudeness, their low price made these very simple three- and four-wheeled cars very popular. Many cycle-car makers closed during World War I or during the economic depression of 1920-21. The rest succumbed to competition from the Austin Seven, which was a properly designed small car built in large enough numbers to offer real motoring at an only slightly greater cost. Within ten years, there were twice as many cars as motorcycles on the road in Great Britain, and the cycle-car had almost disappeared.

Designed by Sir Herbert Austin himself, the Austin Seven was a true car in miniature: it could seat two adults and two children. With the price reduced in 1923 to £165—about forty-six weeks' wages for the average worker in the motor industry—it was soon a success. That price was only about £20 more than the cost of a cycle-car or large motorcycle and sidecar, yet the Austin Seven offered more room and greater comfort in the same amount of road space.

The Austin Seven did not just divert motorcyclists to cars; it also created a new market for small cars. As the pioneering motoring journalist S.F.

Edge said in 1925, "Austin's case was an instance of that very uncommon phenomenon, a supply creating a demand, and filling it to the last ounce and every penny piece."

Until 1924, there had been as many motorcycles in use in Britain as cars—about 400,000 of each. But the 1920's saw vehicle prices fall in real terms by more than fifty percent, as makers tried to increase demand and exploit the cost-savings of mass production. By 1934, there were about 1.3 million cars in use. Cars became twice as common as motorcycles, and the number of motorcycles in use decreased.

The car pictured opposite is the second prototype and differs in many small ways from production cars. It was used for road testing by journalists and to illustrate advertisements and catalogues under the slogan "The Motor for the Millions."

The simple steel chassis was robust, but it was not very rigid. The arrangement of front and rear

LIFE WORTH LIVING.
With an "Austin Seven" you can spend it where you will, for travel costs you about 1d. per mile with your jolly young family. The "Austin Seven" makes happiness inevitable, and it is a smart little turnout to be proud of.

A page from the Austin Seven sales brochure c. 1922, featuring the car shown opposite

axles was a cheap and simple system, but in practice it involved somewhat uncertain handling. Cable-operated brakes were fitted on all four wheels, but they were not very effective. The front brakes were controlled by a hand lever and the rear ones by a pedal. *The Light Car and Cyclecar* magazine reported favorably, "One is impressed by the silence of the axle, transmission and engine in top gear, whilst the suppleness of the suspension and the manner in which the car holds the road on corners are equally remarkable…the steering, too, is finger light…", but they did concede that the brakes were "not startlingly sudden and tremendously powerful in action, but rather retard gradually and progressively." The 696 cubic centimeter engine (increased to 747 cubic centimeters for production cars) was similar in size to those in large motorcycles, but it was built like much larger car engines. Its ten brake-horsepower allowed it to achieve a top speed of about fifty miles per hour. Fuel consumption was around forty-five miles per gallon.

Although electric lights were fitted, this prototype had no dynamo to power them. Electric starters became available on production cars in December 1923, speedometers in February 1924, and shock absorbers in March 1924. The basic design remained in production until 1938, by which time some 291,000 had been produced.

An affectionate, if rather robust, retrospective assessment in *Autocar* magazine in 1967 said, "In view of its haphazard steering and hopeless brakes, I believe that a special providence watched over them, as they bumbled and rumbled and bounced and hopped over the highways of the 1920's." But the Austin Seven did offer a unique balance of comfort, performance and price that created and defined a new small car market.

—PETER MANN

DOBSON SPECTROPHOTOMETER (1924)

Solving the problem of the ozone hole may be considered an environmental success story. Discovered in the mid-1980's by the scientists from the British Antarctic Survey working at the Halley Research Station in Antarctica and quickly attributed to the emission of industrial chemicals known as chlorofluorocarbons (CFCs), the hole has stabilized and may soon be on the mend thanks to the 1987 Montreal Protocol on Substances that Deplete the Ozone Layer, which eventually resulted in a virtual ban of CFC use. The scientists at the Halley station were using a Dobson spectrophotometer to measure the ozone levels. The first such instrument, designed and built in 1924 by Gordon Miller Bourne Dobson (1889–1976), is pictured opposite.

Ozone is a molecule that consists of three oxygen atoms that are joined together. It forms in the stratosphere twenty-two to thirty-one miles above the earth's surface when ultraviolet (UV) light from the sun splits normal oxygen molecules (which are composed of two oxygen atoms) into single atoms; some of these single atoms then combine once more and form ozone. Circulation of air in the atmosphere brings most of the ozone down to about fifteen miles above the earth and disperses it toward the poles. Ozone itself also absorbs UV radiation, and eventually it decomposes back into normal oxygen molecules. Ozone's capacity for absorbing UV light makes the ozone layer a protective filter, preventing from reaching the earth's surface much of the UV radiation that can cause cancers, sunburn and eye damage.

The study of atmospheric ozone was a constant in the brilliant scientific career of G.M.B. Dobson. He studied natural sciences at the University of Cambridge from 1907. He worked at the Kew Observatory, and also as meteorological adviser to the military flying school on Salisbury Plain. In 1920, he became Lecturer in Meteorology at Oxford University, where he began researching atmospheric ozone.

The presence of ozone in the upper atmosphere had been identified in the late 19th century by Walter N. Hartley and Marie Alfred Cornu. In the 1910's, Charles Fabry and Henri Buisson from the University of Aix-Marseille were successful in measuring atmospheric ozone for short periods of time by observing its absorption at various radiation wavelengths. Their apparatus was not, however, suitable for routine outdoor use. Dobson adapted it to produce the more robust and portable Dobson spectrophotometer.

The purpose of the spectrophotometer is to measure the total amount of ozone contained in a vertical column of air extending from the ground to the top of the atmosphere. It works by comparing the intensity of the sun's UV light at two wavelength bands: at one of them, UV light is strongly absorbed by ozone; the other acts as a reference band. The more ozone there is in the air, the greater the proportional difference between the two bands. Incoming sunlight is split by prisms, and the two selected wavelengths are isolated by a device known as a double monochromator. Using optical wedge filters, the operator makes the two signals equal in intensity, and, after calibration, the positions of the optical wedges provide the relative intensities of the two beams, from which the amount of atmospheric ozone can be calculated. Vertical distribution of ozone is determined by measuring upward to different angles when the sun is low. The recording mechanism of the first spectrographs built by Dobson and his colleagues was the use of photographic plates, which had to be sent back to Dobson's Oxford laboratory for processing. In later instruments, the plates were replaced by a photoelectric cell.

In 1925, Dobson used the original prototype to measure seasonal variations of ozone over Oxford. Between 1925 and 1927, he supervised the construction of new instruments, and they were sent to various locations, including Ireland, Switzerland, Germany, California, Egypt, India and New Zealand. By 1956, there were more than forty Dobson spectrophotometers worldwide. The first International Geophysical Year (July 1957 to December 1958) gave a significant boost to the study of atmospheric ozone. This collaborative, international scientific project established a global network of Dobson spectrophotometers, which included what was then called the Halley Bay Research Station. In 1957, to ensure consistent practice across the expanding network, Dobson and his colleagues wrote an article entitled, "Observers' Handbook for the Ozone Spectrophotometer." By 1992, a total of 120 Dobson spectrophotometers had been constructed. Many of them are still in use around the world, alongside spectrophotometers developed in the 1970's by David Wardle and Alan Brewer, as well as satellite-borne instruments. The World Meteorological Organization coordinates this global ozone-observing network.

The most famous chapter in the history of ozone began in 1985, when a group of British scientists (Joseph Farman, Brian Gardiner, and Jonathan Shanklin) published a paper in the journal *Nature*; it postulated that springtime ozone levels over Antarctica were decreasing year by year. Their findings were based on data from a Dobson spectrophotometer, and further examination using satellite data corroborated their findings. Previously, satellite instruments had missed the "ozone hole" because they were programmed to regard extremely low readings as erroneous. In the following years, scientists established that the geographical extent of ozone depletion was increasing, and CFCs were identified as a major cause. The Montreal Protocol went into effect on January 1, 1989; in 2007, a report on ozone depletion by the United Nations Environment Programme concluded that the hole in the ozone layer was the smallest it had been for nearly a decade. (The size of the hole is, however, controlled by fluctuating atmospheric conditions; for example, in 2008, the hole was larger.) Although ozone holes will probably remain in existence until at least 2080, the number of ozone-destroying substances in the atmosphere is now declining. Certainly, G.M.B. Dobson and his spectrophotometer deserve much credit for that.

—VICKY CARROLL

LOGIE BAIRD'S TELEVISION APPARATUS (1926)

In a darkened room in January 1926 a group of men peered intently at a dimly flickering pink-orange oblong of light, about the size of a narrow credit card. Was there anything there? Could they truly see a barely recognizable human head and shoulders, features just discernible in the gloomy changing light? Was the image real or was it imaginary?

The eccentric thirty-eight-year-old Scots inventor, John Logie Baird (1888-1946), presented early observers of television with a very crude picture. Yet, his cumbersome mechanical system, skillfully marketed and publicized, generated a demand for television at a time when many thought it had little to offer.

Like all forms of moving pictures, Baird's system relied on "persistence of vision." This meant that as long as the small parts of a large picture were presented in turn—and quickly enough—the human eye could be fooled into seeing the whole without registering either individual elements or flicker. Thus, a picture, scanned and broken into parts in one place, could be transmitted as an electrical signal to another location where, received and reassembled, could be viewed as a recreated whole.

In his first demonstrations, Baird used a mechanical device to scan his subject, a puppet named Stooky Bill. The heart of the device was a large spinning cardboard disk, containing lenses arranged in a spiral. Baird's version of this idea—first suggested by Paul Nipkow (1860-1940) in Berlin in 1884—contained thirty lenses. As the disk rotated, each lens scanned a different part of the subject. Light from each part of the subject fell, in turn, on to a light-sensitive electrical cell and was converted into an electrical signal. Via the scanning process, the pattern of light and dark in the original image was converted into a stream of electrical signals that could be sent along a wire or transmitted. Baird's system used the then-common medium-wave radio band.

In the receiver, the incoming signal caused a neon light bulb to flicker. A second Nipkow disk, spinning in time with the first, converted the flickers back into lines. The viewers in 1926 saw these lines as a pinkish, glowing picture.

There was growing public interest in this phenomenon. News of the early experiments in the new medium had been published in the popular wireless-enthusiast press of the time. Publicity about Baird's demonstrations—first in local, then in national newspapers—had alerted the general public. Baird's company hoped that the campaign would generate support for its attempt to persuade the Postmaster General to issue a license for experimental public trials and to force the British Broadcasting Corporation to carry such broadcasts.

The campaign succeeded in raising public awareness, though the generally acknowledged poor quality of the images discouraged the BBC and enraged other scientists in the television field, particularly Alan Archibald Campbell-Swinton (1863-1930). He had suggested an electronic television system in 1911, and he objected strongly to what he saw as attempts to defraud the public—both with claims that Baird had invented television and with the suggestion that the mechanical system was the only and/or best solution. Nevertheless, the

John Logie Baird presenting the "transmitting portion of the original experimental Baird television apparatus" to the Science Museum, September 1926

company's persistence paid off. The BBC transmitted experimental television, using the Baird system, from 1929 to 1935. The initial audience was small; Baird himself estimated that no more than thirty sets existed, twenty of which he believed had been built by home enthusiasts.

Amazingly, despite the poor images and the constraints imposed on program makers (head-and-shoulder shots only), public demand increased. To satisfy it, other workers, based in the mainstream electronics industry and dissatisfied with the limitations of mechanical scanners, turned their attention to electronic means of producing pictures. Baird, however, stuck stubbornly to his mechanical system, struggling to improve the quality of the images it produced.

In 1936, the world's first regular high-definition television service began. From the BBC's Alexandra Palace studio, both Baird's mechanical system and the rival electronic system developed by EMI (*see* **Emitron Camera Tube**) were used to broadcast images to the viewing public. Although it involved great improvements over the early apparatus, the still unsatisfactory mechanical system was dropped after three months and the electronic EMI version adopted. Baird had not succeeded.

So how can one consider Baird to be one of the "fathers" of television? It would not be for his making any lasting technical breakthrough, although it might be because he did more than any of his contemporaries to publicize the new medium. He struggled enthusiastically with an ultimately inadequate system while others in Britain, the United States, the Soviet Union and Europe contributed to the development of electronic television. Perhaps his fame also owes something to the public's relish for adoring an appealing underdog who holds that place in the public's heart traditionally given to the gallant loser.

—JANE BYWATERS

PENICILLIN (1928)

The antibiotic penicillin was the miracle drug of the 20th century. The hope provided both to patients and doctors that infections could not just be treated but also quickly cured shaped attitudes to medicine in the years after World War II. Since its wartime introduction, penicillin seems to have epitomized the benefits of science. It has not, however, always been used in a straightforwardly scientific way. It has, instead, sometimes been taken to cure afflictions such as colds and flu, which are caused by viruses not susceptible to antibiotics. One result of this type of questionable use of the drug has been that resistant, hard-to-kill organisms have emerged. Arguably, these practices have spawned a new era: a post-antibiotic age—one in which brand-name, "miracle" drugs become more visible and ostensibly more important than the solid science that went into their creation.

The "branding" of penicillin was achieved partly through a wish to believe and partly through the stories told of it. These stories range from accounts of wonderful transformations experienced by sick and suffering patients, to the history of its discovery by Alexander Fleming (1881–1955), and the drug's isolation, characterization and mass production in World War II. These stories were framed during wartime and the subsequent years of reconstruction to promote national morale, to promote an appreciation of the benefits of modern medicine and to establish commercial reputations.

Although there were many reports of the therapeutic power of molds as far back as Biblical times, Alexander Fleming was the first to describe in detail the material that achieved such benefits and to define which families of bacteria contained the beneficial components. In the autumn of 1928, this highly respected bacteriologist returned to his laboratory bench at London's St. Mary's Hospital Medical School. He found that a plate infected by staphylococci bacteria had also been infected with a penicillin mold—and around the mold, the bacteria had apparently died.

Fleming explored the behavior of the powerful yellow liquid emitted by the mold, but he could not extract the chemical that gave it the properties that had so surprised him. In frustration, he simply called it penicillin. The range of experiments he could conduct with his small quantity of liquid was very limited, and although he did test its effects on eye infections, he was not primarily interested in its therapeutic properties—but rather for its potential as a test for the development of vaccines.

Others in well-staffed, cross-disciplinary laboratories were working on finding cures for infections by means of chemicals. In 1935, a team at Germany's Bayer & Company announced the discovery of Prontosil, the first of the sulfonamide (antibacterial) drugs. These drugs were later shown to work by preventing bacteria from synthesizing vitamin B (folic acid), which is necessary in order for bacteria to reproduce. That fact was not then known, but scientists quickly learned the benefits of this type of drug.

If Fleming's original discovery has often been chalked up to luck and his ability to recognize the unexpected, the next stage in the history of penicillin was the result of advanced science and dogged hard work. At Oxford University, the Australian pathologist Howard Florey (1898-1968), created a team of experts in microbiology, chemistry, and pathology. From 1938, he and the German biochemist, Ernst Chain (1906–79), took an interest in penicillin, and by 1940, Florey had a team of seven scientists and several technicians working on its refinement. At first, the material proved to be a scientific challenge to isolate and characterize; from 1940, it was also a technical challenge to produce and a medical challenge to utilize. British industry had no expertise in mass-producing the products of molds, which needed oxygen to produce (a specialty of laboratories in the Netherlands, Czechoslovakia, Germany, and the United States). As first mice, then human patients demonstrated the benefits of penicillin, the demand for more of the drug increased, and Florey turned to old friends in the United States.

With support from the Rockefeller Foundation, Florey and his colleague, Norman Heatley (1911–2004), flew to the United States in July 1941. Working with those whose expertise lay in the nurture and use of molds, they achieved success in their efforts to have the drug mass-produced. By 1943, significant amounts of penicillin were being made on both sides of the Atlantic, and within a year, American industry was capable of making truly huge amounts.

After scientific and medical success came the trumpeting of achievement amid the gloom of military defeats and jealousy among allies. American industry pointed to its genius at drawing upon what Europeans called their scientific development. In London, the Dean of St. Mary's Hospital Medical School, Lord Moran, promoted his laboratories as the ones that had nurtured penicillin—and therefore the facilities that deserved support in the future. His scientist, Fleming, free of urgent day-to-day wartime duties, proved a convincing image of the fortunate discoverer. Into the postwar years, Fleming would be the very model of the modern scientific hero. In the storytelling about Fleming and penicillin, and about wartime achievement and military benefit, the pharmaceutical brand was born.

The story of penicillin mold was told in *The Life of Sir Alexander Fleming*, a 1959 biography by the French writer, André Maurois. Maurois tells how Fleming attended the first lecture in England about Prontosil, the German sulfonamide drug. Fleming was accompanied by his colleague, the obstetrician Douglas MacLeod. Afterward, Fleming said, "I've got something much better than this, but no one will listen to me." He then presented MacLeod with a sample of the mold, which MacLeod's family later mounted. Perhaps the story of a simple mold can illustrate the transformation of penicillin from a scientific curiosity to a medical blockbuster—the type of blockbuster which seems fascinating, yet commonplace to 21st-century inhabitants, who are exposed to a barrage of new "miracle" drugs every day.

—ROBERT BUD

SUPERMARINE S.6B FLOATPLANE (1931)

On September 13, 1931, the Supermarine S.6B floatplane pictured here won the twelfth Schneider Trophy contest, capturing the last Trophy for Great Britain. Sixteen days later, the same aircraft set the world airspeed record at more than 400 miles per hour.

The Schneider Trophy contest was established in 1912 by Jacques Schneider with the aim of encouraging seaplane development. It was an international competition, with speed trials and tests of seaworthiness. The contest soon became one of the most prestigious of aviation events; as well, it was highly competitive: in its time (between 1913 and 1931), the Trophy passed between teams from France, Great Britain, the United States and Italy. By 1925, the British government was sufficiently convinced of the value of the contest as a stimulus to development that it undertook the support of manufacturers in the design and construction of aircraft and engines for the competition. In subsequent contests, the government also provided pilots, drawn from the High Speed Flight of the Royal Air Force.

The S.6B had its origins in the S.4 streamlined twin-float monoplane designed by R.J. Mitchell (1895-1937) of the Supermarine Aviation Works in Southampton, England for the 1925 Schneider Trophy contest. Unfortunately, this plane crashed during trials, but not before it had revealed the potential of its basic design. This potential was proven in the 1927 contest, in which two S.5 floatplanes won first and second place.

By 1929, the Air Ministry had determined that the Napier Lion engine, which had been fitted to all of the previous Supermarine aircraft entered in the contest, had reached the limit of its development. In the search for another engine, the Ministry chose a twelve-cylinder design from Rolls-Royce, which Henry (later Sir Henry) Royce developed for competitive use. Mitchell designed the S.6 floatplane to house this 1,900 horsepower engine, which went on to win the 1929 contest. Under the rules, a third win would enable Britain to take permanent possession of the Schneider Trophy.

The S.6B S1595 and the British Schneider Trophy team in training at Calshot, August 11, 1931

British participation in the 1931 contest remained in doubt until early that year. Given the prevailing economic depression, the government was unwilling to provide support. In the event, Lady Houston, a wealthy and fervently patriotic widow, gave the £100,000 necessary for a British team to defend the Trophy. By then, there wasn't enough time for Supermarine and Rolls-Royce to produce new designs, so work started on developing the existing airframe and engine as much as possible.

The resulting aircraft was designated S.6B. Like its predecessor, it was an all-metal monoplane mounted on twin floats. Much of the aluminum surface of the aircraft was double-skinned, to provide drag-free cooling radiators for water and oil. Two S.6B aircraft were built to this design: they were given the numbers S1595 and S1596.

In a grueling development program, the power output of the Rolls-Royce "R" engine was raised to 2,330 horsepower, both by increasing engine speed and by improving the supercharging. The connecting rods, crankcase and crankshaft were all redesigned to withstand greater stress, and a special fuel was devised for race conditions.

By September 12, 1931, the date set for the contest, both the French and Italians had withdrawn because of developmental difficulties with their aircraft. After a one-day postponement, the result of bad weather, S1595, piloted by Flight Lieutenant John Boothman, flew round the course

at Calshot (near Southampton) uncontested at an average speed of 340.08 miles per hour to win the Trophy. Later that day, the world airspeed record was raised to 379.05 miles per hour by Flight Lieutenant George Stainforth, who flew S1596. Rolls-Royce was eager to better this record; on September 29th, powered by a "sprint" version of the "R" engine, using another special fuel producing 2,530 horsepower, and with Stainforth at the controls, the S1595 set the world airspeed record at 407.5 miles per hour.

The experience gained by both Supermarine and Rolls-Royce in achieving these results provided important experience for their subsequent development of high-speed aircraft and engines. Mitchell's design of the Spitfire fighter grew directly from his work on the Schneider Trophy aircraft. Similarly, the Rolls-Royce Merlin engine (*see* **The Rolls-Royce Merlin**), the most widely used British airplane engine in World War II, benefited greatly from the development of the "R" engine. Experience with the "R" engine supercharger also provided an important basis for the design of centrifugal compressors, vital both to Merlin engine development and for the jet engine (*see* **Gloster-Whittle E.28/39 Jet Aircraft**). The designers also learned about the importance of fuel quality for improved engine performance; indeed, the use of tetraethyl lead to suppress detonation (or "knock") anticipated the use of 100-octane fuel during the Battle of Britain—which was significant in the performance of the Spitfire and Hurricane fighter aircraft. The indirect contributions of the Schneider Trophy contests to the Allied war effort proved to be of great value.

Late in 1931, S.6B S1595 was placed on display in London's Science Museum, followed in 1948 by the Rolls-Royce "R" "sprint" engine used in the second airspeed-record attempt. Finally, the Schneider Trophy itself arrived in the Museum in 1977, joining the aircraft, the creation of which it had so notably inspired.

—Alex Hayward

LAWRENCE'S ELEVEN-INCH CYCLOTRON (1932)

"I'm going to be famous!" the young experimenter yelled as he ran exuberantly across the university parking lot. Ernest Lawrence had not observed something new; he had conceived a brilliant idea that would, he believed, make it possible to split the atom much more easily than other scientists had dared to believe. Lawrence went on to build his dream—the cyclotron—which did make it possible to split atoms in an apparatus that could be mounted on a laboratory bench.

Lawrence (1901–58) was twenty-seven when he startled his colleagues by moving from Yale to the University of California at Berkeley, then a little-known state university. He soon took up the challenge of probing atomic nuclei, about which little was known except that they measured only about a billionth of a centimeter across. The best hope of investigating nuclei more closely was simply to see what would happen when they were bombarded with *other* nuclei, such as protons (hydrogen nuclei). But, in order to arrange such collisions, these particles had to be accelerated to high energies so that they could actually penetrate the target nuclei. This acceleration can, in principle, be achieved by using a very high electrical voltage to increase the energy of the particles as they move in a straight line. In practice, several hundred thousand volts are necessary, and this can result in severe problems associated with electrical breakdown caused by sparking.

One evening in the spring of 1929, Lawrence came up with an ingenious solution to this problem. He was in the library at Berkeley trying, with his rudimentary knowledge of German, to read an article on particle accelerators, "Über ein neues Prinzip zur Herstellung hoher Spannungen," by Rolf Widerøe. Inspired by what he had read, he wondered whether it would be possible, instead of accelerating them in a straight line, to make particles whirl in a spiral motion, thereby gradually gaining energy.

To test this theory, Lawrence designed a new kind of accelerator. The paths of the particles emerging from the center of the apparatus would be curved by magnets located on both sides of two hollow cavities (later called "dees," based on their resemblance to the letter "D"). By applying voltage across the gap separating the dees, the particles could be accelerated as they cross it for the first time. When the particles next pass the gap, they would be travelling in the opposite direction, but again they could be accelerated by judiciously reversing the direction of the voltage. By continually repeating this process, the operator could accelerate the particles to high energies, ready to be projected toward target nuclei. As Lawrence showed mathematically, the time taken for each half-revolution is the same regardless of the diameter of a particle's orbit, so, remarkably, the reversal of voltage works equally well for all the particles in the apparatus.

Lawrence believed that he had struck gold, but some of his colleagues were skeptical that the idea could be made to work, and this cool reception probably took the wind out of his sails. For some-

The eleven-inch cyclotron in its Berkeley laboratory, c. 1932

one of Lawrence's drive and ambition, it is somewhat surprising that it took him nearly two years to bring his initial plans to fruition. In April 1931, he and his research associate, M. Stanley Livingston (1905–86), reported that they had created the first working cyclotron, which involved a magnet that was four and a half inches in diameter, and had produced protons that had, in effect, been accelerated by 80,000 volts. In his laboratory ten months later, Lawrence was delighted when he saw that his new eleven-inch cyclotron could accelerate protons through more than a million volts.

In his zeal to improve his design, Lawrence had fallen behind his competitors and lost the race to split the atom, which was won by Cockcroft and Walton at the Cavendish Laboratory in Cambridge, England (*see* **Cockcroft and Walton's Accelerator**). In June 1932, they observed an astonishing reaction in which a nucleus was split by an incoming proton into two helium nuclei, a result that Lawrence and his colleagues reproduced three months later. In this case, it was the runner-up who won first prize: Lawrence was awarded the Nobel Prize in Physics in 1939, twelve years before Cockcroft and Walton were awarded theirs.

Cyclotrons were later used to make many experiments to probe the structure of the atomic nucleus and to discover new chemical elements, several of which were discovered at Berkeley. They are now used to produce radioactive materials for use in medical diagnostics and for treating diseases such as cancer. It is in the field of pure science, however, that Lawrence's original idea has had the most applications. His creation is still the basis of the design of today's particle accelerators, which allow scientists to probe the very structure of matter—into regions even smaller than the nucleus. Lawrence's enthusiastic first reaction to his brainstorm was quite right; he had indeed conceived an idea great enough to make him famous.

—Graham Farmelo

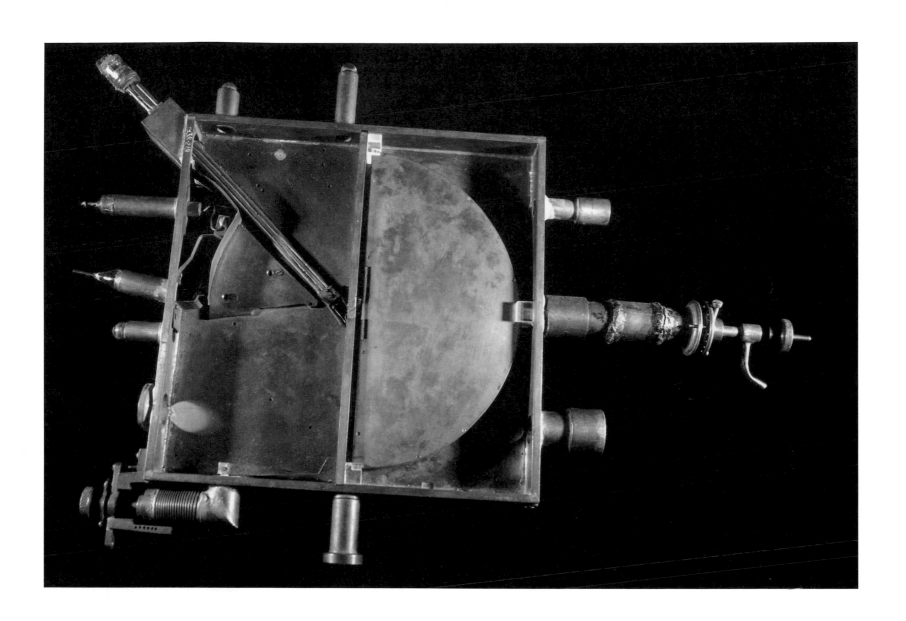

COCKCROFT AND WALTON'S ACCELERATOR (1932)

In the 1930's, nuclear physics became a distinct specialty within physics, the result of the rapid progress being made in understanding the atomic nucleus. Although major discoveries such as the neutron, the positive electron, and artificial radioactivity helped to establish the new field (*see* **Chadwick's Paraffin Wax** and **Discovery of Artificial Radioactivity**), one of the most important factors was the construction and use of new machines for experimenting on nuclei. These machines were known as particle accelerators, or atom-smashers. Two became operational in 1932: one was built by John (later Sir John) Cockcroft (1897-1967) and Ernest T. S. Walton (1903-95) at the Cavendish Laboratory in Cambridge, England; the other was the eleven-inch cyclotron built by Lawrence and Livingston in Berkeley, California (*see* **Lawrence's Eleven-Inch Cyclotron**).

Although these machines differ in important ways, they both needed space and teams of people to operate them, and they both helped to usher in the "big science" that was such a distinctive feature of research after World War II. This trend has continued, and a modern particle accelerator, such as the LHC (Large Hadron Collider) at CERN (European Organization for Nuclear Research), near Geneva, though still incorporating features of these original machines, is exponentially larger physically, and hundreds of people are needed to operate and maintain it.

The accelerator built by Cockcroft and Walton was the first to be used for experiments in nuclear physics, and it helped to keep the Cavendish Laboratory at the forefront of research in nuclear physics in the early 1930's. In this machine, protons, the nuclei of hydrogen atoms, were released at the top of a glass column that had been emptied of air. The protons had a positive electrical charge, and as they traveled down the glass column, they passed through a series of metal cylinders that were electrically charged. The effect of this process was to accelerate the protons. At the far end of the glass tube, placed in the way of the protons, was the tar-

Cockcroft and Walton's accelerator (The Cavendish Laboratory, University of Cambridge)

get: a piece of metal or other material. The protons collided with the nuclei of atoms in the target and broke them into fragments. By examining the fragments, physicists were able to find out more about the structure of these nuclei.

By 1930, physicists were striving to develop accelerators, and at that time they did have the advantage of high-voltage equipment that was being developed by the electrical industry. From before World War I, when Ernest Rutherford (1871–1937) had first suggested that the atom had a central nucleus, most experiments on the nucleus involved radioactive materials. Rutherford, for example, often used polonium as a source of alpha particles (helium nuclei) for bombarding the nuclei in a target. These radioactive materials were, however, both expensive and awkward to

use. In addition, the particles they emitted had a range of energies, and it was difficult to interpret the results. The new accelerators were designed to get past these difficulties by providing beams of particles with only one type of energy.

With Rutherford's encouragement, various scientists at the Cavendish attempted to build accelerators during the 1920's. However, none of them were successful—not until Cockcroft and Walton pooled their resources. To build their machine, they took advantage of some new technology. Before going to Cambridge, Cockcroft had been an apprentice with Metropolitan-Vickers in Manchester. That company made transformers and other items used in the transmission of electricity. That experience, and his continuing contact with Metropolitan-Vickers were very important for Cockcroft when he began work on building a machine to accelerate particles using high voltages. One particular example of the value of this connection was that the glass tubes (originally designed for gasoline pumps) and other components of the accelerator were sealed, using special compounds that had only recently been developed by C.R. Burch, who worked at Metropolitan-Vickers.

The first experiments by Cockcroft and Walton involved bombarding lithium atoms with protons, splitting them to make helium nuclei. For this work they were awarded the Nobel Prize in Physics in 1951. Very quickly, their original machine was superseded by a larger version built by Philips, the Dutch electrical firm. Even in the modern day, the principle of the Cockcroft-Walton machine is often used in the first stage of the largest accelerators: to accelerate the particles so that they can be injected into a circular cavity where they can be given even higher energies. Today's particle accelerators are used to study the subatomic world in the hope of better understanding the matter of the universe we see around us—as well as discovering brand new matter.

—ALAN MORTON

CHADWICK'S PARAFFIN WAX (1932)

The plain pieces of paraffin wax pictured opposite were used by James Chadwick (1891-1974) early in 1932 in experiments that established the existence of neutrons, which are particles found alongside protons in the nuclei of atoms.

Following the announcement of this discovery in the journal *Nature*, a journalist from *The Times* of London reported Chadwick's own assessment of the significance of his discovery as follows: "Positive results in the search for 'neutrons' would add considerably to the existing knowledge on the subject of the construction of matter, and as such would be of the greatest interest to science, but, to humanity in general, the ultimate success or otherwise of the experiments that were being carried out in this direction would make no difference."

Chadwick's off-the-cuff remarks about the significance of his discovery turned out to be remarkably wide of the mark; in the short term, however, his assessment was correct: in 1932, the neutron was important only to a small number of scientists interested in the structure of the nucleus.

Chadwick's own views about nuclear structure were strongly influenced by ideas put forward by Ernest Rutherford a decade or so earlier, shortly after Rutherford had left Manchester to become Cavendish Professor at Cambridge. At that time, it was thought that the atom was made up of a nucleus surrounded by negatively charged electrons. Rutherford speculated that, if the nucleus was made up of positively charged particles (for which he suggested the name protons) as well as extra electrons (in addition to those found outside the nucleus), then it might be possible for a proton to become closely bound to a nuclear electron in a unit he termed the "neutron." Over the next few years, he engaged in experiments at the Cavendish Laboratory to detect these neutrons—without success.

During this time, various scientists made substantial progress in solving problems regarding the behavior of electrons orbiting outside the nucleus. But as quickly as problems were solved for the electrons outside the nucleus, the new theories created problems of description for the electrons inside it. At the end of 1930, in a desperate attempt to solve some of these mysteries concerning nuclear electrons, Wolfgang Pauli suggested another kind of neutron for the nucleus, a particle much lighter than the one proposed by Rutherford.

This theoretical impasse was broken by new experimental evidence. In Germany, France, and Britain, physicists had used alpha particles (helium nuclei) to bombard samples such as beryllium. In these experiments, a nucleus weighing four units was made to collide with one weighing nine units, and the effects were quite dramatic, as well as puzzling—but they did suggest a new effect of radiation. Chadwick, however, did not agree with this conclusion, and he proceeded to carry out his own experiments. He put these pieces of paraffin wax in the way of the mysterious radiation from beryllium. As other experiments had discovered, the radiation knocked hydrogen nuclei (i.e., protons) from the wax, which could easily be detected: they have an electric charge. Chadwick then repeated the experiments using nitrogen; he measured the recoil of the nitrogen nuclei. From his measurements of the energy of the protons and the nitrogen nuclei, Chadwick concluded that they had been hit by particles with the same mass as the proton, but they were electrically neutral—in other words, they were Rutherford's neutrons. This was the view of the neutron that Chadwick put forward in the papers he published announcing his discovery.

Within a few months it was generally accepted by other scientists that nuclei were comprised of protons and neutrons; nuclear electrons went out of favor. Pauli's particle also lost its place as a constituent of the nucleus, but it too was transformed and became a particle created in various nuclear processes. Enrico Fermi (1901–54) suggested the name "neutrino" to distinguish this particle from its heavier namesake.

Chadwick's initial estimate of the value of his discovery was accurate at the time. The ability of neutrons to split uranium nuclei, and in so doing, releasing huge amounts of energy, was discovered during the winter of 1938-39, a few months before the outbreak of World War II. By the end of the war in 1945, and only thirteen years after Chadwick's discovery, neutrons had been used for chain reactions in nuclear reactors and in nuclear weapons. Contrary to Chadwick's initial view, the neutron was a discovery that had a profound effect on all humanity in a remarkably short time.

—ALAN MORTON

*Opposite: The circular paraffin wax samples with pieces
of metal foil also used in Chadwick's experiments*

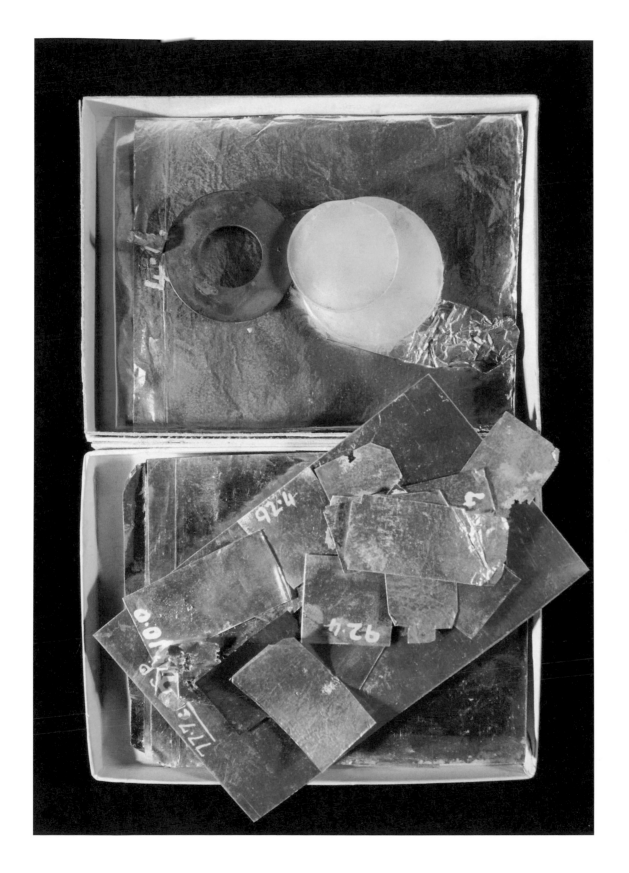

THE DISCOVERY OF POLYETHYLENE (1933)

In 1932, the unambiguously titled book *Chemistry Triumphant* announced that people were destined to live in "The Silicon-Plastic Age." Although silicon was temporarily forgotten, the term "Plastic Age" did prove to be a suitable description for the mid-20th century. Of course, at the time, the new materials were in their infancy. Although flammable celluloid and a selection of rather brittle plastics were available (with Bakelite perhaps being the best remembered; *see* **The First Plastic** and **The Bakelite Coffin**), this new generation of materials was only just starting out. Commercially produced nylon, polystyrene, acrylic glass and polyvinyl chloride (PVC) were all nascent in the 1930's. Indeed, the plastic destined to be made in the largest quantity of all, polyethylene (better known in the U.K. as polythene, Imperial Chemical Industries' name for the material), was not developed until 1933, a year after the "Plastic Age" had been inaugurated. Polyethylene is widely used today, primarily as packaging material.

Brunner, Mond & Company, which with Nobel Industries had taken the lead in forming Imperial Chemical Industries (ICI) in 1926, had become ICI's Alkali Division. Inspired by the tradition of Ludwig Mond, the Alkali Division opened the new Winnington Research Laboratory in 1928. The laboratory recruited many scientists, and embarked on an ambitious program of fundamental research—all this during a worldwide depression. ICI acquired equipment from Anton Michels, a Dutch physicist who had designed devices for subjecting materials to pressures up to 2,000 atmospheres (atm). The company also created a new research program, one that had as its goal the discovery of whether high pressures would produce new and potentially useful effects.

The scientist first assigned in 1931 to this work was Reginald Gibson, who had worked with Michels in Amsterdam; a year later, Gibson was joined by Eric Fawcett, an organic chemist recruited to work on ICI's ambitious plan to process oil from coal (a project that had become a casualty of the Great Depression).

After much fruitless work, on Friday, March 24, 1933, they set up an experiment in which a mixture of ethylene and benzaldehyde was heated to 170 degrees centigrade. The pressure was raised to 1,700 atm, and the apparatus was then left alone, in the hope that the two substances would combine. The experiment had been suggested by Professor (later Sir) Robert Robinson to test the idea that the normally slow Diels-Alder reaction might run rapidly (and without a catalyst) under pressure.

On Monday, no evidence of the desired reaction was found, but Gibson wrote in his notebook,

Apparatus used to discover polyethylene, 1933

"Waxy solid found in the reaction tube." Fawcett then demonstrated that this substance was a polymer of ethylene, which could be melted and drawn into threads.

Attempts to repeat the results of that experiment were generally unsuccessful: either nothing happened, or the reaction decomposed explosively. This early apparatus was amazingly crude; its components consisted of a manually operated pressure pump and a reaction vessel in a bath of hot oil. At the time, the prevailing belief was that it was impossible to make solid polymers of ethylene, so when in September 1935 Fawcett announced at a conference in Cambridge that he had done so, he was greeted with skepticism. That skepticism was fortunate for ICI, as there was no patent application as of yet on file. Only in December of that year, using greatly improved equipment, was Michael Perrin able to carry out the polymerization in a manner that was able to be reproduced, thus establishing the basis for a production process. Following this work, progress was rapid; in 1937, a ten-ton per year pilot plant opened, to be followed by a 100-ton per year plant, which commenced operations on the first day of World War II: September 1, 1939.

Although polyethylene is perhaps most often associated with disposable packaging, its first uses were for insulation, and almost immediately this use became of crucial importance. A 1,000-ton plant that became operational during World War II would provide the lightweight compact insulators that made possible the air- and ship-borne radar that enabled the Allies to fight and win the Battle of the Atlantic (*see* **Randall and Boot's Cavity Magnetron**). No longer was polyethylene just an interesting technological oddity—it was helping to win a war, and was to become perhaps the most ubiquitous plastic in existence.

—ROBERT BUD

THE BOEING 247D (1933)

Occasionally, technological change occurs so rapidly that it becomes inextricably linked with a particular object, and over the years that object comes to stand for both the advance and the social changes associated with it. Just as the rise of railways and the growth in passenger traffic is associated in the minds of many people with Stephenson's *Rocket*, the establishment of regular air routes over much of the globe and the explosive growth in the number of air passengers in the 1930's is automatically linked with the Douglas DC-3. But, upon reflection, perhaps there is another airliner that should be given due consideration for the title of the world's first modern airliner: the sturdy, purposeful Boeing 247D.

In the late 1920's and early 1930's the United States of America was entering a "golden age" of aeronautical design; on the drawing boards of companies such as Boeing and Douglas were studies that would lead to advances such as metal construction, variable-pitch propellers, the pressurized airline cabin, and retractable landing gear. The establishment of government-sponsored airmail routes across the country, initially serviced by ex-World War I biplanes, had led several manufacturers to produce fast, metal monoplanes, and these aircraft were used by the emerging airlines to compete for the lucrative mail contracts. The Boeing Monomail was the company's answer to the problem: it was developed to carry not just 750 pounds of mail but also six passengers in a separate compartment. Eventually, these aerodynamic and structural advantages were brought together in the Boeing 247. This revolutionary aircraft had most of the attributes associated with the modern airplane. The prototype first took to the air in February 1933.

With the ability to climb safely on the power of only one of its two engines, the 247's advanced specifications, including controllable-pitch propellers, retractable landing gear, and wing and tail de-icing, rendered virtually all other airplanes obsolete overnight. Boeing at this time was involved in both aircraft construction and airline management, in that United Air Lines (UAL) and its subsidiaries

were part of the company. This link between producer and consumer (which had been experienced briefly by Handley Page in Great Britain) meant that the developed version of the airplane, the Boeing 247D, was designed for United Air Lines route service. This also meant that, although more than seventy 247D aircraft were eventually built, only UAL used them in quantity. Boeing's withdrawal from airline ownership in 1934 allowed UAL to buy the newer Douglas airplanes, so that by 1938, only twenty-five 247Ds were left in its fleet, while the DC-3 Mainliner was used on its prestigious fifteen-hour coast-to-coast service.

These qualities—reliability and a respectable range—were the primary reasons that one of the Boeing airliners led a rather storied and exciting existence: the NR-257Y, which was selected by Colonel Roscoe Turner, the flamboyant American air racer, and Clyde Pangborn for the 1934 Mac-Robertson air race from Mildenhall in Suffolk, England to Melbourne, Australia. Nursing overheated engines, Turner and Pangborn came in a respectable third place behind the eventual winners, Scott and Black in a de Havilland 88 Comet. It was highly significant that a KLM DC-2, carrying paying passengers, actually came in ahead of the 247D, twenty hours behind the winning Comet. This performance was so impressive that the decline of the

*Boeing 247D N18E arriving at Wroughton,
Wiltshire, England, August 3, 1982*

Boeing 247D and rise of the DC-2/DC-3 line can be traced to this event.

One might assume that the outbreak of World War II would bring a new lease on life for the 247 design, as it did for many other transport aircraft. And, in fact, the 247 design did lend some aerodynamic features to Boeing's famous B-17 Flying Fortress heavy bomber. But the limited number of 247's that were requisitioned by the U.S. Army Air Force suffered from the disadvantage of having main wing spars that penetrated the cabin, and this design generally restricted the plane's usefulness—especially for carrying cargo. The aircraft were declared surplus to requirements in 1944, and they were sold off, mainly to private owners.

Although it had come to the end of its service life, one aircraft of this type was to play a powerful role in the history of aviation. The Royal Air Force had obtained a 247D from the Royal Canadian Air Force, and this aircraft, serial number DZ 203, was used in the development of centimetric radar (*see* **Randall and Boot's Cavity Magnetron**). This 247D's main claim to fame, however, was that it performed a fully automatic approach and landing (one of the world's firsts) on February 10, 1945, at RAF Defford in Worcestershire. The aircraft was flown by Group Captain Frank Griffiths; it was equipped with a modified Minneapolis-Honeywell electronic autopilot. Regrettably, the aircraft was destroyed when a large oak tree was blown down onto the hangar it was occupying at Defford.

The number of Boeing 247Ds still in existence is now very small. The aircraft pictured here (now in the Science Museum) had a fairly unremarkable career with various private owners after World War II, but it became famous on its very last flight. Purchased after the closing of a museum in Florida, and flown across the old World War II Northern Ferry Route, N18E became, when it touched down at Wroughton in Wiltshire on August 3, 1982, the oldest aircraft to have flown across the Atlantic.

—Ross Sharp

DISCOVERY OF ARTIFICIAL RADIOACTIVITY (1934)

Among the remarkable developments in nuclear physics during the early 1930's was the discovery of artificial radioactivity. Appropriately, it was made by Irène Curie (1897–1956) and Frédéric Joliot (1900–58), the daughter and son-in-law of Marie Curie (1867–1934), who had made important contributions to the study of radioactivity at the turn of the 20th century. Marie Curie had studied naturally occurring radioactive materials; the next generation investigated processes by which materials could become radioactive.

In 1932, Curie and Joliot were involved in experiments leading to James Chadwick's discovery of the neutron (see **Chadwick's Paraffin Wax**); they were unlucky enough not to have made the discovery themselves. The pace quickened later that year when Carl Anderson in California found a new positive particle. The evidence for this particle came from photographs of cosmic rays (the debris produced as particles from space travel through the atmosphere) passing through a lead plate. Very soon, Anderson's discovery was linked to a prediction made by Paul Dirac in Cambridge, England of a positively charged version of the electron, and Anderson's particle became known as the positron. These discoveries, together with the work of Curie and Joliot on artificial radioactivity, laid the foundations for nuclear physics, which, by the mid-1930's, had become a distinct specialty within physics.

The significance of these separate discoveries became clear in 1933 when Curie and Joliot continued the experiments that had provided Chadwick with an important clue in his discovery of the neutron. Samples of elements such as boron were bombarded with alpha particles (helium nuclei) emitted by polonium (an element discovered by Marie Curie). But now, Curie and Joliot noticed that positrons were also emitted in some cases. More remarkable was that, when the source of alpha particles was removed so that no more nuclear collisions took place, the positrons were still detected. They had discovered a new nuclear process, analogous to the emission of an electron by the nucleus—which was the process of beta decay discovered at the turn of the century. In beta decay, nuclei do not disintegrate immediately, but only after time; this disintegration varies from nucleus to nucleus because the decay is subject to chance.

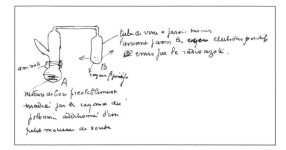

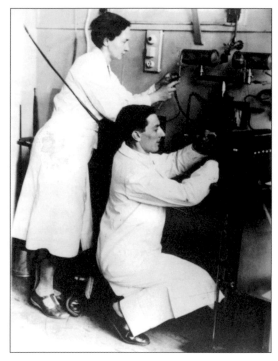

Top: Sketches by Frédéric Joliot describing the apparatus used for producing artificial radioactivity.
Above: Irène Curie and Frédéric Joliot at work in their laboratory, c. 1934 (Institut Curie)

To confirm that they had discovered a new kind of radioactivity, Curie and Joliot hit on an ingenious experiment to prove that it was the nucleus that was involved. They took a small U-shaped tube, filled one arm with a boron compound and sealed the tube. The boron sample was bombarded with alpha particles (helium nuclei) to convert some boron nuclei into nuclei of nitrogen. According to Joliot and Curie, these nitrogen nuclei would now emit positrons. The clever twist to their experiment was to separate this nitrogen from boron or anything else that might be present and to show that it was only nitrogen that emitted positrons. They did this by collecting the nitrogen in the form of ammonia in the other limb of the tube. In the end, all radioactivity had left the boron compound and was carried into the nitrogen. This result showed that the boron had been converted to nitrogen. In 1935, they were awarded the Nobel Prize in Chemistry for their work on this new kind of radioactivity.

Another interesting aspect of the experiment was that Curie and Joliot used electronic methods of counting the positrons they detected. These new methods were developed in the early 1930's and often used components designed for radio receivers.

The discovery of the neutron and of new kinds of radioactivity in the early 1930's paved the way for many new experiments on the properties of different nuclei. Toward the end of the 1930's, Enrico Fermi and others in Rome, and Otto Hahn and Lise Meitner in Berlin carried out experiments bombarding heavy nuclei such as uranium with neutrons. These experiments in turn led to the discovery of nuclear fission a few months before the outbreak of World War II, which then led to the race to build the first atomic weapons. What had been an esoteric branch of science involving a few people such as Marie Curie and Ernest Rutherford, had become, in one generation, important enough to change the course of history.

—ALAN MORTON

REYNOLDS' X-RAY MACHINE (1935)

In the February 11, 1898 issue of the magazine *The English Mechanic and World of Science*, readers saw an article explaining how London general practitioner John Reynolds and his fifteen-year-old son Russell (1880–1964) had constructed at their home a spark induction coil. This device, effectively a step-up transformer, was essentially one of the very early x-ray machines.

Although it is somewhat difficult to imagine an adolescent building a working example of the latest medical imaging machine (even if the child's father is a medical doctor), it was indeed possible to do in 1896, the year after a then little-known German professor, Wilhelm Conrad Röntgen (1845-1923) first described x-rays. All the parts of the Reynolds apparatus were more or less standard components that would have been in stock at the time at any supplier of physics equipment—or capable of being made at home by a good amateur technician.

Physicists—notably William Crookes, a friend of John Reynolds—had been experimenting for some years with discharge tubes—glass vessels with a partial vacuum and electrodes sealed into each end, and which glowed when high voltages were applied. The first x-rays were produced using Crookes' tubes, and the Reynolds machine would most likely have featured this kind of tube in its first incarnation. Reynolds added the 1900-vintage tube pictured opposite—the "Russell Reynolds Adjustable Electrode"—to the original machine in 1950; he had donated the machine to London's Science Museum in 1938.

This apparatus, typical of those early years, was comprised of batteries not unlike car batteries. It contained an "interrupter" to convert the current of the batteries from direct current to the alternating current suitable for the induction coil. As well, it contained the tube itself. It was completed by a fluorescent screen (on which an x-ray image could be seen in a darkened room, without the need for a photographic plate). Like the tube, the screen pictured opposite was added later; it also dates from about 1950.

Ten years after his initial x-ray set, Russell Reynolds qualified as a doctor, and from 1922 he devoted his medical research to the challenging problem of moving-image x-ray film—or "cineradiography." The technical problems he faced were enormous. From about 1900, those trying to make cineradiographic films had used one of two methods. The "direct" method recorded x-ray images directly onto cine-film. X-rays could not be focused, so film extents were huge. To record adult chest movements, a strip of film fifteen inches wide and 120 feet long was required. Other researchers, including Reynolds, pursued the "indirect" method, which entailed pointing cine-cameras at x-ray images on fluorescent screens and photographing those images. The main problem with this method, however, was obtaining an image that was bright enough. As well, the high doses of radiation that were required were potentially harmful to the patient and often put an intolerable strain on the x-ray tube. By 1925, Reynolds had partially solved the problem by incorporating a switch that made the x-ray tube operational only when the cine-camera shutter was open, thus reducing the risk to both patient and tube. By 1935, he had made films of almost all human joints at speeds of up to twelve frames per second, and was able to convince Watson & Sons (the leading British manufacturer of radiology equipment) that the method was potentially useful enough to warrant commercial production. Cautiously, Watsons put together much of the first commercial cineradiography apparatus by modifying existing equipment; the camera-base, for example, is a modified dental chair.

Cineradiography was never designed to be a money-making enterprise, though research interest in the technique continued throughout the 1940's and 1950's. Reynolds himself succeeded in producing cine-film of moving joints at up to fifty frames per second; he had created slow-motion films (which were essential to understanding the heartbeat) by projecting film at twenty-five frames per second. Barium-meal examinations of the intestines were also recorded. Reynolds achieved these later successes not by any single breakthrough, but by constant modification in technique: using wider aperture lenses, more brilliant fluorescent screens, and more sensitive film—as each became available.

In 1951, Watsons introduced an improved version of Reynolds' latest apparatus, and it was exhibited at the 1951 Festival of Britain as an outstanding example of scientific equipment. Only four were produced, with three going to hospitals and one to a research institution. Cineradiography was actually incorporated into routine medical practice a decade later, with the application of television technology—the electronic amplification of which reduced the need for large doses of harmful x-rays. Watching and recording moving x-ray images on monitors is now an integral part of complex medical procedures (such as cardiac catheterization)—a practice that began in part due to an adolescent's creation.

—TIM BOON

ORIGINAL RADAR RECEIVER (1935)

In England on February 26, 1935, Robert Watson-Watt (1892–1973; knighted in 1942) carried out an experiment that demonstrated that a distant aircraft could reflect radio waves transmitted toward it, and that sufficiently sensitive equipment with a cathode-ray tube could detect the radio echo.

The previous day, equipment from the Radio Research Station near Slough was loaded into an old Morris van and driven to a field at Weedon, Northamptonshire. The equipment was duly prepared, and the next day Watson-Watt arrived with A.P. Rowe (assistant to H.E. Wimperis, Director of Scientific Research to the Air Ministry), to join his own assistant, A.F. Wilkins. Secrecy surrounded the proceedings; those not directly involved with the experiment had to wait outside the field. The pilot of a Heyford bomber had been ordered to fly at 5,000 feet, flying both toward and away from the powerful beam from the BBC's nearby Empire shortwave transmitter at Daventry. Convincing deflections of the spot on the screen were seen when the aircraft was some eight miles away, and Rowe, now satisfied that the aircraft could be detected by radio waves, made his way back to London in Watson-Watt's car to report back to the Ministry.

At this time, Watson-Watt was Superintendent of the Radio Division of the National Physical Laboratory. His work was concerned with understanding upper-atmosphere reflection—the basis of global radio communication in the pre-satellite era. Since Marconi's 1901 transatlantic demonstration, the power of radio for communication, entertainment, and propaganda had spawned a great deal of investigation into the way that radio waves followed the curvature of the earth. In England, Sir Edward Appleton pioneered this work in 1924, pointing a radio beam skyward and analyzing what came back; he was honored by having one of the ionized layers of the upper atmosphere named for him.

One of the standard tools of the Radio Research Station's investigations was a crude two-channel radio receiver, linked to a cathode-ray tube that indicated the strength of reflections. It was one of these instruments that was used for the now celebrated Daventry Experiment. Watson-Watt's motive in conducting the experiment and showing it to A.P. Rowe dated back to a request that he had received from the Air Ministry in 1934 to investigate the feasibility of a "death-ray" method of ensuring that bombers did *not* get through to

Top: Receiver room of a Chain Home radar station, England, 1939-45 (Imperial War Museum).
Above: Sir Robert Watson-Watt at the presentation of the "Daventry Experiment" receiver to the Science Museum in 1957

their targets. His first calculations suggested that a lethal beam was not practicable, but late in January 1935, one of his memoranda concluded, "Meanwhile attention is being turned to the still difficult but less unpromising problem of radio-detection as opposed to radio-destruction…"

Only weeks later (and before the important experiment on February 26th), Watson-Watt proposed the linking of radio-detection stations to a fighter-control network. On the basis of the Daventry demonstration, and the Air Ministry's reading of the worsening European situation, the development of coastal defense radar in England, linked to a system of fighter control, went ahead remarkably rapidly and paved the way for the British (and later, Allied) radar superiority that was to be a decisive factor in victory during World War II. Indeed, Watson-Watt's central role both in the development of radar and in turning British radar from an experiment in an old van into a vital instrument of defense is undeniable.

In 1951, Watson-Watt and some of his colleagues (who had formed a syndicate) became embroiled in an extraordinary dispute with the Royal Commission on Awards to Inventors. During the proceedings, the syndicate suggested that "three-pence per head in respect of himself and of each of those dependants who are not able to speak for themselves" might not be an unreasonable sum to seek—yielding about £675,000. The actual award, however, was £87,950. Watson-Watt's powers of persuasion failed in more lighthearted circumstances in 1954 in Canada. On his way to a learned gathering where he was to speak about his work on the development of radar, he was stopped for speeding by a highway patrol officer who was using an early radar speed trap. The officer remained unmoved by Watson-Watt's claims and demanded payment of the fine on the spot.

—Eryl Davies

MANCHESTER DIFFERENTIAL ANALYZER (1935)

Many problems in science and technology can be described by differential equations; in the 1920's and the 1930's, early analog mechanical computers, differential analyzers, were built to solve them. Modern digital computers are essentially counting devices that can handle numbers, but these differential analyzers solved equations by modeling mathematical relations with shafts and gears.

The reconstructed section, pictured opposite, of the University of Manchester's differential analyzer of 1935 is one of the surviving relics of this era in computing. It is about half the size of the complete machine, built by the local engineering firm of Metropolitan-Vickers under the supervision of Professor D.R. Hartree (1897-1958). He first used it to solve problems in atomic physics, and during World War II it was used for military research, including work on the cavity magnetron (*see* **Randall and Boot's Cavity Magnetron**).

Its design was based on the first differential analyzer, completed in 1930 at the Massachusetts Institute of Technology by a team led by Vannevar Bush (1890-1974), an electrical engineer who later marshaled America's scientific effort in World War II; he also played a key role in the decision to drop the first atomic bomb.

The drive to develop the differential analyzer came from the needs of the growing telecommunications and electricity supply networks of the 1920's. To design power transmission networks that would not be liable to blackouts, engineers had to solve complex differential equations which could take months of painstaking work. Bush realized that

"better ways of analyzing were certainly needed." His solution, like that of Charles Babbage a hundred years before, was to mechanize the process of calculation (*see* **Babbage's Calculating Engines**). In the 1920's, the only feasible way of making such a computer was to use mechanical rather than electrical technology. Relatively complex analog mechanical computing devices had been developed to aim the guns aboard battleships, but Bush's machine was the first civilian analog mechanical computer. The idea of the differential analyzer had been put forward by the British physicist and entrepreneur Lord Kelvin in 1876 (he did not attempt to build one), but Bush's team did not know of Kelvin's ideas until after their differential analyzer had been completed.

The MIT differential analyzer was widely copied in Europe as well as in America. It was the best means of solving differential equations until the development of the electronic computer during the 1940's. During World War II, differential analyzers were chiefly used to compute firing tables for guns, as well as to design radar control systems and electronic circuits. Setting up a differential analyzer could, however, take several days. Although hybrid—part mechanical, part electrical—other differential analyzers were built during the war to speed up the process. Ultimately, however, they were a dead-end technology, made obsolete by digital computers.

Hartree wrote that "my first impression on seeing the photographs of Dr. Bush's machine was that they looked as if someone had been enjoying himself with a super-Meccano [a predecessor to the

Erector] set." Hartree began trying to build a Meccano model "more for amusement than with any serious purpose," which was so successful that, with the help of a student, Arthur Porter, he built a small differential analyzer, using many standard Meccano parts. It was capable of useful work, and it offered good practice in programming while the full-size apparatus was under construction.

The differential analyzer was a purely mechanical device—a sort of analog computer that worked by representing each element of the equation by a rotating shaft. Gearing allowed the shafts to be set to rotate relative to each other, reproducing the relationship of elements of the equation. Experienced operators could follow the mathematics of solving an equation by watching the turning shafts.

The first step in using a differential analyzer was to set up the shafts to reproduce the equation to be solved, a time-consuming operation. Graphs were prepared showing how changing variables in the equation altered with respect to one another; for example, in the case of a train's speed changing with time. These graphs were placed on input tables, and, as the analyzer ran, the curves were followed by human operators who moved pointers geared to the mechanism, feeding in the varying values.

Output tables on the differential analyzer produced graphs showing the solution to the equations that had been set up beforehand. In our sample example, the output would be a graph of distance travelled by the train versus time.

—PETER TURVEY

EMITRON CAMERA TUBE (1935)

HMV, Columbia and the Great Depression seem an unlikely combination to advance the story of television. But in the 1930's, these protagonists joined the scattered assortment of inventors and scientists working worldwide to perfect the new medium, and they developed a television camera that was to be at the forefront of the field for many years.

By 1930, the British recording company HMV was well established as a manufacturer of record players and records, and it had assembled a very strong team to research new methods of recording sound for domestic use and for the cinema. The company's interest in film led naturally to experiments with electronic pictures, and a team led by W.F. Tedham (1907-39) began to look into the possibilities of both mechanical and electronic television.

The Columbia Graphophone Company was HMV's biggest rival, and much of its success at this time was due to the efforts of a young electronics genius, Alan Blumlein (1903-42), who, under the leadership of Isaac Shoenberg (1880-1963), had invented a system for recording gramophone records. This system, which later allowed for the production of stereo records, allowed the company to avoid paying the crippling patent royalties faced by other companies.

In 1931, as the Great Depression began to take its toll, sales of both record players and records plunged. In order to survive, the two companies were forced to merge. The new company was called EMI (Electric and Musical Industries Ltd.), and the combined research team under Shoenberg was one of the most powerful collections of industrial research brainpower ever gathered in Britain. As the new company took stock of the depressed marketplace, it realized that an outstanding new product was necessary to ensure survival. The research team was instructed to develop a viable television receiver.

In order to test the new electronic receivers, EMI also built a mechanical camera, using technology already demonstrated by Baird (*see* **Logie Baird's Television Apparatus**). Two of the team, Tedham and James Dwyer McGee (1903-87), argued strongly that the company should also research the next logical step: the development of an all-electronic tube. Their request was, however, turned down, and the pair was instructed not to waste their time on this project.

Tedham and McGee were so convinced that their tube could be made to work that they secretly constructed one in the autumn of 1932. They borrowed the equipment to test it, and after some minor adjustments, the picture appeared as if by magic on the receiver. It surprised them both so much that they simply watched in fascination until the picture slowly faded as the tube failed, making no effort to record the achievement with a photograph. They now knew that a functioning tube like the one they had constructed was possible, but as they had defied management instructions to build it, they quietly put the now defunct tube away in a cupboard. They did, however, continue to argue for the approval to develop the tube.

In 1933, Vladimir Zworykin (1889-1982), working for RCA in the United States, published an account of his electronic camera tube (the iconoscope), which had given promising results. Despite the lack of detail in the description, it was obvious to Tedham and McGee that Zworykin's tube was basically the same as their own experimental tube of the previous year. Armed with this information, they were at long last able to convince EMI management to authorize research into an electronic camera tube.

Despite a deteriorating financial situation, EMI expanded its research department, and by the end of 1933, more than sixty people were working on the development of television. McGee was able to demonstrate presentable pictures from a series of experimental tubes by the beginning of the following year, and he instilled enough confidence in the company's directors for them to cease development of mechanical cameras and concentrate on all-electronic television.

When the world's first regular high-definition television service started from London's Alexandra Palace in 1936, the three cameras in the EMI studio had at their heart the Emitron camera tube. Its task was to convert the light from the scene into a television signal. The way in which it did so was essentially the same as in the experimental tube made by Tedham and McGee four years earlier. Although it initially produced pictures that looked no better than those of its rival, the Emitron had a number of very significant advantages, including the ability to produce good pictures in daylight (its rival, the Baird mechanical system, required powerful studio lighting). Although the reliability of the Emitron camera was not good by today's standards, it was markedly better than Baird's system. Indeed, many of the early Emitron cameras were still in service nearly twenty years later.

The development of the Emitron enabled Britain to take the lead in television during this brief but critical pre-war period, and it laid the foundation for the country's later preeminence in television broadcasting.

The original Emitron camera tube was essentially the same as the iconoscope tube made by RCA in the United States, and in the years since then there have been several claims that the British tube was a straight copy of the American version. Although the extent to which the two companies collaborated is still unclear, some evidence indicates that the two tubes were developed completely independently. The technology of the day dictated that such a tube could be made in only one way, and that could be what accounts for the similarity between the two.

—John Trenouth

MALLARD (1938)

"Britain's Fastest Ever Express Does 125 Miles an Hour"—thus did London's *Daily Mail* newspaper announce the running of the high-speed streamlined steam locomotive *Mallard* on Sunday, July 3, 1938. It is now accepted that *Mallard* briefly touched 126 miles per hour and had established the highest authenticated speed ever achieved by a steam locomotive. After running 1,426,621 revenue-earning miles, *Mallard* was withdrawn from service in April 1963.

If *Mallard* looks fast and exciting even today, it is a tribute to a design that astonished and excited observers when the first of the A4-class locomotives (of which *Mallard* was one) appeared in 1935. A total of thirty-five A4 locomotives were built between 1935 and 1938.

The streamlined A4 was a dramatic and effective response by Sir Nigel Gresley (1876-1941), Chief Engineer of the London and North Eastern Railway, to an increasing public demand for faster express passenger trains. All over the world, between 1930 and the outbreak of war in 1939, there was a sudden and astonishing increase in railway speeds. In Germany in 1933, the diesel-powered *Fliegende Hamburger* streamlined train was running daily between Hamburg and Berlin at an average speed of 77.4 miles per hour. In the United States, larger diesel streamliners like the *Pioneer Zephyr* were providing high-speed luxury travel.

After observing German diesel traction and French high-speed gasoline railcars, Gresley put his faith in steam. He developed and refined his earlier express passenger designs, of which the A1-class *Flying Scotsman* is probably the most famous. What distinguished the A4 was the streamlined casing that hid a conventional steam-locomotive shape under a smooth exterior. Streamlining both reduced wind resistance and provided an eminently marketable shape—one with noticeable style and panache. After he had conducted exhaustive wind-tunnel tests, Gresley adopted a horizontal-wedge shape for the streamlining similar to that used by Ettore Bugatti on his 1923 Grand Prix cars and subsequently in 1933 on a fleet of Bugatti's railcars,

Mallard just prior to its record-breaking run on Sunday, July 3, 1938 at Barkston on the East Coast Main Line

which Gresley had observed in France. The wedge shape helped with lifting smoke away from the driver's field of vision; it also reduced air resistance to the locomotive itself and avoided disturbance to passing trains.

The exterior of the A4-class locomotive was visibly different from that of the A3 design before it, but, to the engineer, the changes beneath the streamlining were even more significant. The working pressure of the boiler in the A3 was 220 pounds per square inch; in the A4, it was raised to 250 pounds per square inch. Gresley paid careful attention to the size and shape of the steam and exhaust passages and to their internal streamlining. He reduced the cylinder diameter on the A4 by half an inch to eighteen and a half inches, and he increased the piston-valve diameter to nine inches in order to give maximum ease of steam flow between the valves and cylinders. Inside and out, the A4 locomotive was built for speed. The first of the class, *Silver Link*, ran some 10,000 miles in its first three weeks of service, of which 7,000 miles were at an average speed of more than seventy miles per hour.

Mallard left Doncaster works, where it was built, in March 1938. The locomotive was different from its predecessors in that it had a double chimney and blast pipe, which improved the exhaust of steam

from the cylinders. Proof of the effectiveness of the system and of the quality of engineering that created *Mallard* was the 126 miles-per-hour record achieved only a few months after the engine first ran.

It is the world speed record that gives *Mallard* its special distinction. In the hands of driver Joe Duddington and fireman Tommy Bray, with inspector Sid Jenkins riding on the footplate, *Mallard* was ostensibly taking part in one of a series of high-speed braking tests. For the test on July 3, 1938, Gresley had decided, in great secrecy, to attempt to beat the speed record of 114 miles per hour then held by the London, Midland and Scottish Railway. With a train of seven vehicles, including a dynamometer car to record and measure the test, *Mallard* ran from King's Cross station in London to just north of Grantham, where it was reversed on a triangle of track. The speed record was achieved on the return trip, as the train ran down Stoke Bank between Grantham and Peterborough.

The calculations from the dynamometer record established (as the daily papers triumphantly claimed the next day) 125 miles per hour over a distance of a quarter of a mile. Although this fact was not mentioned in the press at a time when Nazi Germany was already threatening world peace, the Deutsche Reichsbahn (the German national railway) record of 124.5 miles per hour was just bettered. Subsequently, a more detailed study of the *Mallard* dynamometer roll showed that, for a few brief seconds, the train had achieved 126 miles per hour.

"…Once over the top I gave *Mallard* her head and she just jumped to it like a live thing. After three miles the speedometer in my cab showed 107 miles an hour, then 108, 109, 110 …'Go on old girl,' I thought, we can do better than this. So I nursed her and shot through Little Bytham at 123 and in the next one and a quarter miles the needle crept up further—123-½ – 124 – 125 – and then for a quarter of a mile…126 miles an hour…"—Driver Joe Duddington, speaking on the BBC Home Service, April 1944.

—Rob Shorland-Ball

THE BAKELITE COFFIN (1938)

Plastic—a contradiction, a mystery, a tireless facto-tum—a triumph of creative chemistry.—John Mumford, *The Story of Bakelite*, 1924.

In 1938, an Englishman, James Doleman, designed a Bakelite coffin—at the time the largest compression molding ever made. Doleman was unable to capitalize on his invention; he died in World War II. His compression molding machine was taken over by the British War Office to make military components. At its Eagle Works Bakelite factory at Stalybridge, near Manchester, the Ultralite Casket Company Ltd. reputedly made only five of these coffins, which contained molding powder supplied by Bakelite Ltd. of London.

The world may not have been ready for plastic coffins in 1938, but they are more common today; perhaps it was a design ahead of its time. The British design elite and the public were notoriously resistant to non-natural materials; they were more comfortable with plastics that resembled natural materials—such as celluloid that looked like ivory or tortoiseshell, or cast phenol imitating jade or amber. The Bakelite coffin is a good example of the contradiction of Bakelite: it is made to look like wood—indeed, mahogany—yet it is made of the first truly synthetic plastic.

Bakelite is a brand name for phenol-formaldehyde resins and related products. It was first developed in America in 1907 by the eponymous Dr. Leo Baekeland (1863–1944), a brilliant Belgian-born chemist and a self-made man who combined academic excellence with entrepreneurial skills. Bakelite is a thermosetting polymer made by combining phenol and formaldehyde, products of the coal-tar industry.

Baekeland was in competition with a contemporary Scottish engineer, Sir James Swinburne. They built on the work of former pioneers in the field of phenolic resin, including Kleeberg, Story, and Luft. Baekeland fervently wrote in his diary: "I have an excellent thing, and it would be a great disappointment if my patent had been preceded by an earlier invention of somebody else." (July 14, 1907)

Baekeland triumphed—by one day. The development of Bakelite was not, however, a eureka moment; rather, it was the culmination of five years of detailed, persistent, and focused work in his laboratory.

Dr. Baekeland combined persistence and enormous self-belief with business skills. A life-threatening disease—a ruptured appendix—gave Baekeland the imperative to focus. His subsequent successful invention, Velox photographic paper, which he sold to Eastman Kodak in 1899 for $750,000, made him independently wealthy and enabled him to pursue whatever avenues of research he pleased.

He worked in a number of industries, notably in rubber and electrochemical. These experiences gave him a useful understanding of various manufacturing processes; they also showed him what these burgeoning new industries could offer in terms of new products. Baekeland then decided to concentrate on a replacement for shellac. For this, he used an exudate of the lac beetle to create a varnish; combined with other materials (such as wood), it also made a semi-synthetic moldable material popular for decorative goods.

Baekeland's key contribution to the production of a commercially viable material based on phenol-formaldehyde resin was in recognizing the need to use heat and pressure. His work culminated in his so called "heat and pressure" patent, submitted in 1907 and granted in 1909. He then developed the bakelizer, a giant autoclave in which he was to produce what later commentators have spoken of as his "magical" material.

Once Baekeland had produced his new material, he became an energetic promoter of it. He received a rapturous reception for his invention on February 5, 1909, at the New York chapter of the American Chemical Society. His first customer was the Boonton Rubber Company. Early products of Baekeland's Bakelite Corporation exploited Bakelite's excellent insulating properties, and included electrical components, parts for the emerging automotive industry, and, very quickly, military components—which became an important and large percentage of his product base. The more consumer-oriented products arrived later, from 1924 onward, when his public relations manager, Allan Brown, recognized the existence of a large female consumer market, and enthusiastically advertised Bakelite products to a wider audience. It quickly became known as the material of a thousand uses, and by 1928, Baekeland had appeared on the front cover of *Time* magazine.

In Britain, a Bakelite company was set up in 1927, when Dr. Leo Baekeland and Sir James Swinburne signed an agreement between the Bakelite Corporation and The Damard Lacquer Company, forming Bakelite Ltd. The Bakelite coffin is perhaps its most unusual legacy.

—Sue Mossman

RANDALL AND BOOT'S CAVITY MAGNETRON (1940)

In 1940, the cavity magnetron was at the top of the list of Great Britain's secrets. Its military possibilities were such that there were grave doubts about the wisdom of telling even Britain's closest allies about it. It was the key to smaller, more agile radar sets having vastly improved detection capability. On February 21, 1940, a cavity magnetron developed by J.T. Randall (1905–84; knighted in 1962) and H.A.H. Boot (1917–83) worked for the first time in a laboratory in the Physics Department of the University of Birmingham. It was the culmination of an urgent search for a high-power source of ultra-short radio waves—those with wavelengths measured in centimeters.

In order to advance radar technology, the far-sighted British Admiralty had, in 1939, placed a contract with the University of Birmingham to develop centimeter-wave generators; the task was overseen by the head of the Physics Department, the Australian Professor Mark Oliphant. He provided resourceful leadership and cushioned his scientists from the unfamiliar problems of military security and wartime cutbacks, providing them with an environment where creativity could flourish.

Radar development in England had begun in earnest in 1935 (*see* **Original Radar Receiver**). The Chain Home system devised by Robert Watson-Watt's team operated at wavelengths of several meters. A basic rule of radio engineering states that an aerial system, in order to form a directional beam of radio waves, must have dimensions that are several times greater than the wavelength used. Consequently, the aerials used for the fixed Chain Home radar stations measured more than one hundred meters in height and width. Clearly, more easily steerable (and mobile) radars had to have smaller aerials; equally, if accuracy and definition were not to be lost, shorter wavelengths had to

be used. Radar engineers (including German ones) knew that centimetric (rather than metric) wavelengths would be best.

A fundamental problem with valves operating at short wavelengths is that the time taken for electrons to travel between the internal parts, the electrodes, is significant compared to the rate at which the radio signals are fluctuating in the external circuits to which they are connected. The cavity magnetron was developed by Randall and Boot to bypass this problem and to generate short wavelengths reliably. Mathematical analysis of the electronic paths in the cavity magnetron was one of the more complex and important tasks undertaken by D.R. Hartree's group, using the Manchester differential analyzer (*see* **Manchester Differential Analyzer**).

Original cavity magnetron, 1940

At first, it was difficult to measure the power and the wavelength of the radio waves generated by the cavity magnetrons. The waves could make light-bulb filaments explode, and it was possible to light a cigarette. The consensus was that, at the very least, the power of a few hundred watts was available. Soon, it was established that the wavelength was 9.8 centimeters; they had aimed for ten. Some bold individuals in the laboratory found that the output of the cavity magnetron warmed up their fingers—presaging an important peacetime role.

Nearly three quarters of a century after those days of unbridled invention, the cavity magnetron finds itself in more commonplace environments. One is at the heart of every microwave oven, producing the radio waves that heat the food. Despite fears of harmful radiation emanating from them, the microwave oven (which first appeared in the mid-1940's) is still very much a part of 21st-century domestic life.

The cavity magnetron transformed allied radar techniques. By 1941, ships were fitted with centimetric radars that could often detect submarine periscopes as they rose a few feet above water. Soon afterward, aircraft were equipped with cavity magnetron-based radars that improved the detection of other aircraft and provided ground-mapping systems for use at night and in clouds. "Blind-fire" gunnery also benefited from centimetric radar. And although the wartime application was the inventors' immediate reward, they could hardly have even dreamed of the widespread peacetime applications of radar that followed after 1945.

—ERYL DAVIES

THE ROLLS-ROYCE MERLIN (1940)

The Rolls-Royce Merlin is uniquely famous. Perhaps the only aircraft engine to be widely known by name, it represented in its time a pinnacle of engineering design and production skill. During the Battle of Britain, the Merlin powered both types of the most significant defending fighter aircraft, namely the Hurricane and the Spitfire. In 1940, few were aware that the Merlin came from a long line of Rolls-Royce engines bearing names of birds of prey. Then the resonances became Arthurian, speaking of magic and power—an engineering magic that only Rolls-Royce could have created.

Rolls-Royce had been drawn into the aircraft engine business at the government's request during World War I. At first, Britain had almost no aircraft engine industry, and planes were equipped with imported French engines or with engines license-built from French designs. Rolls-Royce undertook the manufacture of some of these types, but the company's engineers regarded them as defective. Henry Royce therefore began work on an engine of his own that came to be known as the "Eagle."

The Eagle was a 200-horsepower water-cooled engine with twelve cylinders in a "V" formation. Shortly after World War I, the dependability of these Rolls-Royce engines was demonstrated in a dramatic way by the first crossing of the Atlantic (*see* **Alcock and Brown's Vickers Vimy**) by an aircraft powered by two Eagles.

The experience of designing the Eagle passed into the ethos of the company. The directors even compiled into a book the letters and memorandum from Royce regarding development of the engine, *The First Aero Engine Made by Rolls-Royce Limited*, "as an example to all grades by Rolls-Royce Engineers, present and future."

In the interwar period, the British Air Ministry sustained three aircraft engine firms, sometimes referred to as "The Family," by a considered rationing of contracts. Bristol manufactured high-powered air-cooled radial engines, favored for the emerging airlines; Armstrong-Siddeley also manufactured radials, though of less-developed design, which were used in Royal Air Force training aircraft and in many military aircraft sold for export; Rolls-Royce built water-cooled in-line engines, chosen by aircraft designers for the clean aerodynamic nose they preferred on high-speed fighters.

By the late 1920's, the Eagle had been replaced by the Kestrel—a 21-liter V-12 unit that provided nearly 500 horsepower—which was fitted into the most modern fighters flown by the Royal Air Force. Oddly, in 1935, it was a Kestrel that Germany used for the first flights of both prototypes for the Messerschmitt Bf 109 and for the Junkers Ju 87 Stuka, before suitable German engines were ready.

In this period, Rolls-Royce also produced the 2,000 horsepower "R" racing engines, which achieved notable success in the Schneider Trophy races (*see* **Supermarine S.6B Floatplane**).

The Merlin, fruit of all this piston-engine experience, was designed in 1932. It was intended to develop 750 horsepower, with the possibility of further development to create 1,000 horsepower.

The Merlin engine in the Spitfire exhibition at the Science Museum, London, 2005

Again, it used the V-12 layout and had a capacity of twenty-seven liters. It first flew in 1935. Later that year, a Merlin powered the prototype Hawker Hurricane for its first flight, and in March 1936, it powered the first Spitfire. Throughout the ensuing war, the Merlin was to undergo extraordinary development. Its power more than doubled over five years, mainly through ever-improving supercharger design; it reached well over 2,000 horsepower. The Merlin was also the subject of a major industrial effort: more than 150,000 were being produced by Rolls-Royce at Derby, Crewe and Glasgow, by the Ford Motor Company in Manchester, and by Packard in the United States.

But the Merlin will always be best remembered for its role in the Battle of Britain, for this was as much a contest between engines as between aircraft. No other engine made anywhere in the world at the time could have given the Spitfires and Hurricanes the power and stamina to meet the Daimler-Benz-powered Messerschmitts.

Both at the time and now, looking back more than seventy years, one sees that the Battle of Britain was one of the pivotal military contests in human history. It was the first check inflicted on German military might during World War II, and as a result, the United Kingdom survived to be the base for the Anglo-American invasion of Normandy in 1944. It might be said that modern Europe was born in 1940 over the fields of Kent. After the war, Lord Tedder, Marshal of the Royal Air Force, who had been in charge of the development of aircraft and engines during the Battle of Britain, attributed the British victory to three predominant factors: the skill and bravery of the pilots, 100-octane fuel and the Rolls-Royce Merlin engine. There can hardly be a more powerful testimony to the social and human impact of airplane engineering.

—ANDREW NAHUM

GLOSTER-WHITTLE E.28/39 JET AIRCRAFT (1941)

On April 8, 1941, the first British jet aircraft lifted off briefly while on taxiing trials and flew straight and level for about 200 yards. To the surprise of the Power Jets engineers who had built the W.1 engine, the flying men seemed greatly relieved. Jet propulsion was so new that pilots had privately wondered whether the high-speed jet exhaust would make the aircraft squirm around uncontrollably, "like a dropped garden hose."

After the trials, the aircraft was taken back, but not to the Gloster works where it had been built. Because there was a fear of sabotage, it was taken to the premises of a provincial motor dealer, Crabtree's Garage in Cheltenham. The uniqueness of the occasion was compounded when the pilot filled in the test flight report and noted under the entry for type of propeller, "no airscrew necessary with this method of propulsion."

The first proper flight took place some weeks later, on May 15th, at Cranwell in Lincolnshire. The aircraft flew for some seventeen minutes, and the most notable thing about this and the ensuing flights was the almost complete absence of trouble from the new engine. These flights gave a powerful boost to the view that the Whittle gas-turbine jet program was a "potential war winner."

Frank Whittle (1907–96) was a cadet at the Royal Air Force College, Cranwell, when the idea for jet propulsion first came to him as he was writing a thesis on "Future Developments in Aircraft Design" as part of his class work. Although he also reviewed the possible use of the gas turbine in the thesis, he had not put these two ideas together and was thinking in terms of a jet produced by a piston engine buried within a hollow fuselage and driving a fan. He concluded that this configuration would not offer any advantage over a conventional propeller, but in the following year, 1929, he realized that a gas turbine could be used to generate a propulsive jet.

Whittle communicated his ideas to the Air Ministry, but the Ministry's feeling at the time was that materials were not available to stand the high temperatures in the turbine and that the develop-ment of such an engine, consequently, would be exceedingly difficult and expensive. Whittle was posted to test-flying duties at Felixstowe, where he made numerous experimental catapult takeoffs; he was commended as "a very keen young officer and a useful test pilot." In this period, he continued to think about the jet engine and tried to interest various companies in his ideas. He wasn't successful.

His outstanding performance in the Officers' Engineering course persuaded the Air Ministry to send Whittle to Cambridge, where he studied mechanical science. While at Cambridge, he allowed his patents on his engine to lapse, but in 1935 he received a letter from a former Royal Air Force colleague, Rolf Dudley Williams, c/o General Enterprises Ltd., Callard House, Regent Street, London. It said: "This is just a hurried note to tell you that I have just met a man who is a bit of a big noise in an engineering concern and to whom I mentioned your invention of an aeroplane, *sans* propeller as it were, and who is very interested…. Do give this your earnest consideration and even if you can't do anything about the above you might have something else that is good."

General Enterprises, the unlikely springboard for the British turbojet revolution, was a company making and supplying cigarette vending machines,

Above: Gloster-Whittle E.28/39 in flight, c. 1940's.
Opposite: The W.1 engine used for the first flights

which Williams ran with another ex-RAF man, J.C.B. Tinling. It was the latter's father, a consulting engineer, who encouraged them to "get into the aircraft business" to be ready for the impending war.

The enthusiasm of Williams and Tinling helped bring about the creation of Power Jets Ltd. to develop Whittle's ideas. They found financing from a firm of City of London bankers, and the Air Ministry agreed to let Whittle work as chief engineer for the company. The British Thomson-Houston Company, manufacturers of steam turbines, was subcontracted to build much of the engine, and initially the tests took place in the BTH plant at Rugby.

By April 1937, the first ground test engine, known as the WU (Whittle Unit), was running, although the problem of obtaining controlled stable combustion led to its being redesigned twice. By the summer of 1939, the engine was approaching its designated thrust. At this stage, the Air Ministry became enthusiastic about the work of Whittle's team, and design work for the Gloster-Whittle E.28/39 was begun.

In spite of the success of this aircraft and its engine, the subsequent wartime development of the gas turbine was far from smooth. The decision was made to put the engine into production, but the design was altered to provide more thrust, and this modified version consisted of unproven elements. Further, the relationship was poor between the Whittle team at Power Jets and the Rover car company, which had been selected to manufacture the engine. Matters improved only when Rolls-Royce took over the project, and the Gloster Meteor, the first British jet fighter, entered service in July 1944 in time to be used in action against the German V-1 "buzz bombs."

In 1944, Power Jets Ltd. was nationalized, eventually making up part of the National Gas Turbine Establishment (NGTE) at Farnborough. The tiny company had only had nine years of life, but it had helped to set the stage for an enormous change in aviation.

—ANDREW NAHUM

THE V-2 ROCKET (c. 1945)

On July 20, 1969, Neil Armstrong and Buzz Aldrin became the first humans to set foot on the moon when they landed in the Sea of Tranquility. They were launched on their voyage of exploration by the massive Saturn V rocket. Weighing 3,000 tons at launch, this vehicle had been developed in the United States by a team of engineers led by Wernher von Braun (1912–77). Some thirty years earlier, in Germany, von Braun and many of his Saturn V team had been working on a completely different rocket, which would turn out to be the world's first long-range ballistic missile, the first space rocket, and the forerunner of all modern space rockets. This was the German A-4 long-range bombardment missile, later to be dubbed the V-2 (Vergeltungswaffe 2, or "vengeance weapon") by Nazi propagandists.

Interest in rocketry had surged in the late 1920's in both Europe and the United States, and many amateur rocket societies came into being. Little official interest was shown in their activities, and their members were seen as impractical (and occasionally dangerous) enthusiasts. The situation in Germany was different. Germany's armed forces had been limited as to what they could or could not do by the 1919 Versailles Treaty, and its leaders were looking for ways to bypass these restrictions. The German Army ordnance department had appointed an engineer, Walter Dornberger, to investigate the rocket as a possible weapon of war, and in 1932, Dornberger contacted the Society for Space Travel, the VfR (Verein für Raumschiffahrt), which was then a group that conducted experiments with rockets.

Although he was unimpressed by their technical work, Dornberger paid attention to the enthusiasm of one of the young members, Wernher von Braun, and in November of 1932, von Braun joined Dornberger at the army's rocket test site at Kummersdorf, twenty-five kilometers from Berlin. Von Braun began work on a series of experimental liquid-fueled rockets that would, in just ten years, lead from the early A-1 and A-2 rockets (with thrusts of 300 kilograms and capable of heights of a few kilometers) to the A-4 rocket (with twenty-five tons of

V-2 rocket on launch pad, Operation Backfire, Cuxhaven, Germany, 1945

thrust and that could fly to a height of more than a hundred kilometers). The later rockets were far too large to be developed at Kummersdorf, and the government moved the establishment to a secret site at Peenemünde on Germany's Baltic coast. The move was completed in 1940, and on October 3, 1942, the first successful A-4 launch took place. This rocket reached an altitude of ninety-six kilometers and splashed down on target, 192 kilometers away in the Baltic.

The dimensions of the A-4 had been fixed by the size of the railway tunnels in Germany, and Dornberger decided, rather arbitrarily, that the rocket ought to have twice the range and more than ten times the payload of the German long-range Paris Gun (Paris-Geschütz) of World War I. The V-2 missile was the result: it was 14.3 meters tall, made of mild steel, and it weighed 12.8 tons at launch. Powered by a twenty-five-ton thrust liquid oxygen/alcohol motor, it could carry a one-ton payload over a range of about 300 kilometers. Its flight was controlled by four movable carbon vanes in the rocket's exhaust, and four and a half seconds after launch, the rocket's guidance system began to tilt the rocket on to its preprogrammed flight path. After about sixty seconds of powered flight, when the rocket had gained the velocity necessary to reach the target (some 5,500 kilometers per hour), the motor cut off, and the V-2 then continued

flying on a free-fall ballistic path to its target.

As the first space rocket, capable of flying more than a hundred kilometers into the earth's atmosphere, the V-2 was a milestone on the path to a moon landing. But its effectiveness during World War II as a long-range missile is not so easily assessed. The V-2 offensive lasted from only September 1944 to March 1945. About 6,250 V-2s were produced, and of these 1,115 reached Britain, half of them striking London and causing 2,700 deaths and 6,500 serious injuries. A further 1,775 V-2s hit targets on the Continent (such as Antwerp), and 2,100 were in field storage at the end of the war. The remaining rockets were either launch failures or test rounds.

Most V-2s were made at Mittelwerk, an underground factory in central Germany (near Nordhausen) where, at its peak, the factory produced 700 every month. Forced labor from the nearby Dora-Mittelbau concentration camp was employed at Mittelwerk, and an estimated 20,000 inmates died while working there. But, despite this effort and sacrifice, the V-2 came too late in the war to have any effect on its outcome; just the same, its potential as a potent new weapon was widely recognized by the victors.

A squabble over securing the secrets of German rocketry occurred at the end of the war. The best engineering talent, including von Braun and Dornberger (along with enough material to make about a hundred V-2s), went to the United States. Peenemünde and Mittelwerk were in the Soviet zone of occupation, and some important engineers and a supply of V-2s went to the Soviet Union. Both countries eventually developed the V-2 into the intercontinental ballistic missiles and the satellite launch vehicles that are known today.

Britain gained V-2 experience from the launching (rather than the receiving) end in October 1945 when, with the help of German technicians, three V-2s were fired from Cuxhaven during "Operation Backfire." The V-2 pictured opposite was one of the unused Backfire rockets.

—JOHN BECKLAKE

THE FIRST MARINE GAS TURBINE (1947)

The advent of steam propulsion early in the 19th century and the subsequent growth of scientific tank-testing of hull shapes based on mathematical analysis confirmed one of the laws of physics that governs a ship's behavior in water. Shipbuilders had suspected it for a long time. What John Scott Russell, William Froude (*see* **Swan and Raven**), and their contemporaries were able to investigate, using improved experimental methods, was the relationship between increases in the power available to drive a ship and increases in that ship's speed. Anyone who has rowed a dinghy or paddled a canoe will testify that a doubling of muscular effort does not produce a doubling in speed. A fifty-horsepower engine that propels a fishing boat at a maximum speed of eight knots would have to be replaced by one of 169-horsepower to achieve a fifty percent increase in speed to twelve knots.

Naval architects had long been preoccupied with this phenomenon, since victory in a naval engagement would often go to the force that could outpace its enemy, maneuver itself into the most favorable attacking position, and retire quickly before the enemy had time to retaliate. But, experience showed that a warship typically spent only ten percent of its operational time at full speed; for the vast majority of the time, the output required of its main engines would have been much less. To meet this combination of requirements, for cruising power, naval engineers have tended to rely on proven engine designs, known to be both reliable and economical in fuel consumption, supplemented where necessary by high-performance machinery, the great fuel consumption of which is acceptable if it can provide the extra speed that is crucially important during engagements.

An early and innovative response to this requirement was the marine steam turbine first demonstrated at sea by Charles Parsons in 1894 (*see* **Parsons' Marine Steam Turbine**). Parsons' brilliantly simple invention rapidly found favor with the major navies of the world, and derivatives of it continue to propel nuclear submarines more than a century later. Improvements in boiler performance early in the 20th century led to substantially increased power outputs from marine steam turbines, and in turn these better turbines prompted improvements in hull design, which enabled fast warships to remain at sea in weather conditions that were less than ideal.

Experimental work during the 1930's had established the very favorable power/weight ratio offered by the gas turbine. As the name implies, this later variant used the energy of gas, which was produced by the combustion of fuel in an adjacent chamber. Because a heavy, huge steam boiler was not needed, the gas turbine offered excellent compactness, together with a rapid availability of full power from a cold start. In Britain, by the time the nation's first jet aircraft had flown successfully (*see* **Gloster-Whittle E.28/39 Jet Aircraft**), the Admiralty was already beginning to consider applications of the same principle to warship propulsion.

In August 1943, the Metropolitan-Vickers Electrical Company was commissioned to provide a variant of its gas-turbine engine for an auxiliary power plant in a warship. With industrial production already overwhelmed by the need to produce armaments and munitions, both manufacturing capacity and materials were in grievously short supply in Britain. To save time on unnecessary new development work, Metropolitan-Vickers used an existing F2/3 jet engine as a gas generator. Instead of an exhaust cone, the engine was fitted with a power turbine driven by the exhaust gases, coupled to a propeller shaft via a reduction gearbox. The engineers involved in this project were working at the limits of material science, and the high-performance alloys essential to this power plant were in very short supply. The war in Europe had already ended by the time all of the technical and supply problems had been overcome.

Early in 1946, an existing motor gunboat, the *MGB 2009*, was selected from the Coastal Forces fleet in order to be fitted with the experimental engine, serially numbered G1. Of the existing three Packard gasoline engines in its engine room, the center one was removed and replaced by the Metropolitan-Vickers gas turbine. Although rated at twice the power of the Packard engine it replaced, the gas turbine occupied no more space. A newly designed propeller was required to absorb the 2,500 horsepower on the center shaft, and elaborate fire precautions were necessary to prevent fumes from igniting in the bilges when the gas turbine started up. This was achieved largely by airtight ducting.

By coincidence, the trials of the G1 engine in *MGB 2009* took place in the Solent (the strait separating the Isle of Wight from the English mainland) in July 1947, almost exactly fifty years after Charles Parsons had demonstrated his *Turbinia* at the Diamond Jubilee Review in the same waters. A comparison of the particulars of each vessel follows:

	1897 TURBINIA	1947 MGB 2009
Overall length:	100 feet	116 feet
Beam:	9 feet	20 feet
Displacement:	44 tons	100 tons
Power:	2,100 shaft horsepower	4,050 shaft horsepower

The speed achieved in each case was thirty-four knots.

In 1953, the fast patrol boats *Bold Pioneer* and *Bold Pathfinder* emerged as the first naval vessels to be built specifically for gas-turbine propulsion, each employing two Metropolitan-Vickers G2 turbines. Today, virtually all medium and large warships rely on some version of the gas turbine to supplement their steam turbine or, more normally, diesel engines when full speed is required. Improvements in fuel efficiency have enhanced the reputation of the gas turbine as a compact power plant that can deliver full power less than a minute after being started. Such capability will always be important to naval architects and to the sailors who are responsible for bringing their ships home safely.

—JOHN ROBINSON

VITAMIN B12 MODEL (1947)

In 1885, Thomas Addison, of Guy's Hospital, London, noted the existence of "idiopathic anemia," now better known as pernicious anemia. George Minot, an assistant professor of medicine at Harvard, made the first breakthrough in the treatment of this hitherto fatal disease in the early 1920's when he fed patients with raw liver. Realizing that there was something in the liver that was combating the disease, he asked Edwin Cohn, professor of physical chemistry at Harvard, for his help with its extraction. Subsequently, Eli Lilly and Co. marketed Liver Extract 343 in 1928.

The problem of having to use relapsed pernicious anemia patients to determine the biological activity of extracts was solved by Mary Shorb at the University of Maryland; she developed a microbiological assay method in 1947. The development of liquid-liquid partition chromatography by Archer Martin and Richard Synge in 1941 greatly assisted the extraction, not least because colored bands could be followed down the chromatography column. As a result, Lester Smith's team at Glaxo was able to show that the active factor was in the pinkish-red band. On December 11, 1947, Edward L. Rickes of the pharmaceutical firm Merck was the first to isolate the deep-red crystals of the vitamin. Glaxo announced its own success in the isolation of B12 at a meeting of Britain's Biochemical Society on May 29, 1948. These achievements in themselves, however, did not make vitamin B12 more readily available. That was made possible by Karl Folkers' group at Merck: they discovered in the same year that the microorganism *Streptomyces griseus* could produce B12 when it was fed with a cobalt supplement. This breakthrough led to a fall in the price of the vitamin from $325/gram in 1952

to $8/gram in 1968, which in turn led to its routine use in animal feed.

Several teams of chemists attempted to work out the structure of B12, including teams from Merck, Glaxo, British Drug Houses, and the team of Alexander Todd at Cambridge University. But the structure of vitamin B12 was so complex that chemists made only limited progress over the next eight years. Before the development of nuclear magnetic resonance (NMR), x-ray crystallography was the only way to determine the complete structure. Lester Smith (from Glaxo) knew Dorothy Crowfoot Hodgkin (1910–94) as a result of their service together on the wartime Penicillin Committee, and he gave her the first crystal at the Biochemical Society meeting in May 1948. Glaxo subsequently supplied better crystals, as well as the information that vitamin B12 contained a cobalt atom. Although the task was daunting even for an experienced crystallographer, Hodgkin was aided by the existence of a cobalt atom in the center of the molecule, which allowed her to use the reliable "heavy atom" method (*see* **Festival of Britain Patterns**); as well, she profited from the introductory use of computers (such as the Mark I at the University of Manchester) in the field of crystallography in the early 1950's. She was also assisted by Alexander Todd, who switched his focus from trying to solve the problem of the complete B12 structure to doing work that would solve puzzles that evolved from just the crystallographic work.

Hodgkin published the full structure of vitamin B12 in 1957, probably the first time that a complex chemical structure had been determined by x-ray crystallography. (DNA, although much larger, has a relatively simple chemical structure.) The founder

of modern x-ray crystallography, Lawrence Bragg, described the feat as "breaking the sound barrier" in chemistry. It established Hodgkin's standing as the leading x-ray crystallographer of her generation and led directly to her 1964 Nobel Prize in Chemistry. She then went on to determine the complete structure of the hormone insulin in 1969.

As soon as the structure of vitamin B12 was published, Robert Burns Woodward (1917–79) of Harvard, generally considered the best synthetic organic chemist in the world at that time, began the synthesis of this complex molecule. He later joined forces with Albert Eschenmoser of Zurich Polytechnic, and their joint team synthesized cobryic acid on March 21, 1973. As cobryic acid had already been converted into cyanocobalamin (the chemical name of vitamin B12), there was no need to make vitamin B12 to claim the total synthesis. Nevertheless, Woodward pressed on, and, with his student, Mark Wuonola, he finally completed total synthesis on March 17, 1976. Together, x-ray crystallographers and organic chemists had conquered one of nature's most complex molecules.

Dorothy Hodgkin originally constructed the model pictured opposite for the 1958 World's Fair in Brussels (famous for the Atomium). The model was restored by a molecular model-maker in Cambridge under Hodgkin's supervision in order to ensure that her latest findings were incorporated into the structure. Part of the model is featured on a British postage stamp honoring Dorothy Hodgkin, part of a series of stamps celebrating "women of achievement" issued by the Royal Mail in 1996.

—Peter Morris

PILOT AUTOMATIC COMPUTING ENGINE (1950)

The pilot Automatic Computing Engine (ACE) was one of Britain's earliest stored-program computers, and it is the oldest complete general-purpose electronic computer in Britain. It was designed and built during 1949 and 1950 at the National Physical Laboratory (NPL) at Teddington in Middlesex (Greater London).

The design of this computer came from an earlier and larger computer designed by Alan Turing (1912–54) at the NPL between autumn 1945 and autumn 1947. Turing produced seven different logic designs for the ACE: ACE Marks I-V were designed and tested by Turing between autumn 1945 and spring 1946; ACE Marks VI-VII were developed thereafter with help from James Wilkinson and Mike Woodger, two mathematicians at the NPL. The original ACE models were essentially pen-and-paper exercises in mathematical logic. Though the machines were never built, Turing and his team would write and test programs on each design, using Turing's mathematical proofs and running each program through the intended logical operations bit by bit, instruction by instruction.

Turing had previous experience in designing the logic for computing devices; he had worked on the top-secret "Colossus" code-breaking machines during World War II at the British government's Bletchley Park center. He also had extensive contact with other computer pioneers working in England and the United States, including John von Neumann (1903–57), who pioneered the basic form of computer architecture that is still used in most computers today.

The National Physical Laboratory's original plan was to use outside contractors to construct the ACE. In 1947, however, an electronics group was set up at the NPL under the direction of H.A. Thomas, and it was decided that this group would construct the ACE. Thomas was interested in industrial electronics; he did not want to share any of his precious resources with a mathematics-based project. The ensuing infighting caused months of delays and problems, which eventually led to the departure of both Turing and the head of the mathematics team in the autumn of 1947.

In 1947, the ACE team was expanded to include Donald Davies, an electronics expert, and Gerald Alway, a mathematician. Another new recruit was Harry Huskey, who joined the team in January 1947 while waiting to take a position at the University of Oklahoma. Huskey had worked on the technical description of the University of Pennsylvania's ENIAC (Electronic Numerical Integrator Analyzer and Computer), the world's first electronic digital computer. Huskey managed to persuade everyone except Turing that the only way forward would be to build a smaller version of the Mark V ACE design that could function as a pilot for Turing's ideas.

Work on the test assembly started in mid-1947, but progress was slow, because only a few members of the team had any electronic-circuit design knowledge. By the time that Huskey left in January 1948, most of the design work was complete, but no actual construction had started. In 1948, F.M. Colebrook replaced Thomas as the head of the electronics division. Colebrook was more sympathetic toward the ACE team, and he suggested that they join the electronics team so that both could

The pilot ACE computer at the time of its first public demonstration in 1950

work together. The new unified team worked well, and by late 1948, plans were in place to produce the pilot form of the ACE—a system that was part Mark V and part test assembly.

Construction started in early 1948. The main chassis was completed by December, the control chassis by the following January, and, by February 1950, there were enough working modules for the system to be capable of running some limited programs. The first full run was on May 10, 1950; the program switched the front panel lights on and off in succession. The simple test program consisted of a few instructions, which were entered by hand on the pilot ACE's thirty-two front-panel switches—because the punched card input/output unit had not been fully implemented.

A three-day press demonstration was held in late November 1950, and the pilot ACE worked perfectly throughout. However, the ACE then failed to work again for the next several weeks. Over the following six months, the team changed the internal circuitry and more delay lines were added. The next time the pilot ACE ran was on June 26, 1951, when at three in the afternoon it printed out simultaneously the solution to seventeen equations which had been inputted earlier that morning.

The pilot ACE was estimated to have cost £50,000 (equal to well over £1,000,000 today) to design and build. That cost was recouped many times over. In 1954 alone it earned more than £24,000 on work that included computing bomb trajectories, calculating aircraft wing flutter, and testing crystallography theory.

The machine continued to operate at the NPL until June 1956 when it was finally retired. The *Daily Mirror* newspaper reported this event with the headline, "The Government's First Robot Brain Gets the Sack." The reported reasons were "because faster robots are in use, and this one is so old and tired it's making mistakes."

—Marcus Austin

FESTIVAL OF BRITAIN PATTERNS (1951)

The 1951 Festival of Britain, held on the South Bank of the River Thames in London, was intended to capture and exhibit all that was great about postwar Britain. As it also celebrated the centenary of the Great Exhibition of 1851, which was widely regarded as the forerunner of Britain's modern scientific industry, the concept of the progress of science in the intervening century played a central role both in the Dome of Discovery on the South Bank as well as in the Exhibition of Science in the Science Museum. Although the awesome power of nuclear weapons was on everyone's mind (Britain would explode its first atomic bomb a few months later, in 1952), the nation was a leader in another aspect of atoms: x-ray crystallography.

Ever since John Dalton first used his wooden atoms to try to explain the structure of molecules (*see* **Dalton's Wooden Atoms**), chemists had struggled to discover how molecules were constructed. A key breakthrough, just before World War I, had been the use of x-rays by the father-and-son team of William Henry Bragg and William Lawrence Bragg to probe the internal arrangements of atoms in a crystal lattice. The British school of x-ray crystallography was then developed by Lawrence Bragg at the Royal Institution in London and by the young Irish physicist John Desmond ("Sage") Bernal at Cambridge University. Several female x-ray crystallographers, including Kathleen Lonsdale and Dorothy Crowfoot Hodgkin, were also working on its development—at a time when women were still uncommon in the field. The great breakthrough came in 1935, with the introduction of a heavy metal atom (which was easily detected by x-rays) by the Scottish x-ray crystallographer J. Monteath Robertson. Using this "heavy atom" method, there were several x-ray crystallographic triumphs in Britain in the early postwar years—including the first determination of a steroid compound (cholesteryl iodide) by C.H. Carlisle and Dorothy Crowfoot Hodgkin, and the structure of the antibiotic penicillin by Hodgkin and Charles Bunn.

But, there was the question of exactly how these achievements could be celebrated at the Festival of Britain. Chemical structures were usually presented in the form of colored balls and sticks, but the structures of several of the compounds then being studied by crystallographers were as yet undetermined (for example, hemoglobin), so the models were often difficult to understand. Although "atomic" balls and sticks were an important design motif at the Festival ("atomic" coat-hangers and magazine racks became popular household features), another method of representing x-ray crystallography had to be found.

When x-rays are shone on a crystal, the diffraction of the x-rays, as they pass through the crystal lattice, produces a characteristic pattern on photographic film. Until computers were available to improve the process in the late 1950's, crystallographers had to decipher manually the arrangement of atoms in the lattice by viewing these patterns on film. The Cambridge crystallographer Helen Megaw came up with the idea of displaying these patterns on fabric. This innovation had the advantage of allowing many different patterns to be displayed—which could be interpreted on two levels: as genuine diffraction patterns created by scientists, and as just as intricate but symmetrical designs created by designers and the general public. Once her idea had been accepted, Megaw sought suitable diffraction patterns from her fellow crystallographers, including Bragg, Hodgkin, Lonsdale, Robertson, Bunn, Gordon Cox and Max Perutz. Transferred to fabric, paper, ceramics and glass, these patterns were used in the Festival's Regatta Restaurant, elsewhere on the South Bank, and at the Exhibition of Science in the Science Museum. Several leading British manufacturers also became involved as a result of the formation of the Festival Pattern Group, which included Imperial Chemical Industries, Wedgwood, Dunlop, Chance Brothers and General Electric—as well as stalwarts of 1950's design such as the wallpaper company John Line & Sons and Old Bleach Linen.

The patterns were marketed in the form of fabric, crockery, glass (both drinking vessels and window glass), and stationery, but the concept did not gain wide acceptance—unlike textile designer Lucienne Day's celebrated Calyx design made for the Festival's Home and Garden pavilion. Perhaps the idea was too scientific for public taste, and perhaps it was perceived as being too "masculine" (despite having been created by a woman) at a time when most designs for the home were chosen by women. In recent years, however, with a general rise in critical and public interest in 1950's design, there has been a revival of interest in the Festival of Britain patterns, to the extent that an exhibition of these patterns was held at the Wellcome Collection in London during the summer of 2008.

—Peter Morris

George Farmiloe & Sons, LTD
34 ST. JOHN ST., LONDON, E.C.I.
Sample of
Leathered

OSMAN
FABRICS
Quality
Design/Shade
Width *FABRIC*
Length *C*
BARLOW & JONES LTD
MANCHESTER

NUCLEAR MAGNETIC RESONANCE (1951–53)

Because some atoms (to be precise: their nuclei), and in particular, hydrogen, can be tiny magnets, they emit radio signals when a radio beam is shone on a small spinning sample of the material being studied. For the technique of nuclear magnetic resonance (NMR) to work, however, all these tiny magnets must be lined up in the same direction, like compass needles in the earth's magnetic field. This goal is achieved by placing the sample in a strong magnetic field. The NMR effect was first reported in 1946 by Felix Bloch at Stanford University and, independently, by Edward Purcell at Harvard University. James Arnold, a graduate student at Stanford, obtained the first NMR spectra in 1951. They showed separate resonances for hydrogen nuclei (protons) located at different positions in the molecule. It was now possible to determine the structure of chemical compounds (initially, simple compounds such as ethanol) by studying the spectrum. Herbert Gutowsky, at the University of Illinois, who studied spin-spin coupling, was one of the first to introduce NMR into organic chemistry in the early 1950's. While he was stirring a cup of tea in 1954, Bloch had the clever idea of using a spinning tube to reduce variations in the samples. The magnet for the NMR apparatus pictured opposite was the first magnet in Britain constructed especially for this new technique. It was built by Rex (later Sir Rex) Richards (born 1922) and his colleagues in an Oxford University chemistry laboratory in the early 1950's, and it contained more than thirty kilometers of wire for the hand-wound coils of the electromagnet.

The first commercial NMR spectrometer was marketed in 1952 by Varian Associates, which was one of the first high-tech companies to set up in Silicon Valley, California. The spectrometer, operating at 30 MHz, was purchased by large companies, but it was too expensive for academic use. In the late 1950's, the use of NMR in organic chemistry was heavily promoted by Varian. A pioneer of NMR, Jim Shoolery, was employed by the firm to visit organic chemistry departments to show how the device could be used to reveal the structures of complex chemical compounds; at the same time, he was carrying out his own research on the structure of steroids. Routine use of NMR, however, became possible only when an affordable machine was put on the market. The breakthrough came with the introduction of the A-60 by Varian in 1961 and the advent of a new generation of organic chemists with a stronger grasp of physics than their predecessors.

The interpretation of NMR spectra was greatly improved by the growing availability of stronger magnetic fields, which produced a greater separation between the peaks in the spectrum. By the end of the 1960's, both the powerful superconducting magnets and the more sophisticated Fourier Transform instruments were commercially available. Frank Anet and Tony Bourn introduced the Nuclear Overhauser Effect (discovered in 1953) into chemical NMR in 1965. It was useful for the study of the shape of the molecule in three dimensions (as in the case of sugars) because it provided information about the positions of the protons in space. The veteran British NMR physicist Raymond Andrew developed the technique of "magic-angle" spinning in 1971. By the 1970's, NMR was being used to determine the structure of nucleic acids, proteins and even enzymes. The 1980's saw a second wave of innovations in NMR involving the study of NMR spectra in two and three dimensions, using a combination of proton and carbon-13 NMR for structural work.

NMR has become greatly popular in chemistry not only because it is a highly sensitive probe into the molecule itself, but also because it is so versatile. Comparing the results of proton and carbon-13 NMR allows the entire molecular structure of a compound to be determined quickly from a tiny sample, and without the necessity of crystallization or the use of other techniques. By contrast, x-ray crystallography requires a crystal and considerable analysis of the resulting photographs before the structure can be determined. With the development of two- and three-dimensional NMR, much of molecular biology has also been conquered. NMR is used to study the conformations of large proteins with the aim of understanding how they function in the body and, perhaps even more importantly, how they malfunction—such as they do in medical conditions like cystic fibrosis and Alzheimer's disease.

NMR has evolved from being an esoteric discipline studied in a few university chemistry departments in the early 1950's to being a routine but crucial service discipline in most universities and pharmaceutical companies today. Thanks to the development of NMR by chemists such as Rex Richards, the determination of the molecular structure of complex organic compounds and biomolecules has, for a chemist or x-ray crystallographer, been reduced from a lifetime's work to the work of a few short weeks.

—PETER MORRIS

Opposite: NMR electromagnet built by Rex Richards and colleagues

CRICK AND WATSON'S DNA MODEL (1953)

"We wish to suggest a structure for the salt of deoxyribose nucleic acid (D.N.A.). This structure has novel features which are of considerable biological interest." So began the letter to the journal *Nature* in the April 25, 1953 issue, announcing one of the most far-reaching discoveries of the 20th century: solving the structure of DNA, the substance of which genes are made.

Biotechnology and genetic engineering, which developed so dramatically in the latter part of the 20th century, owe their origins to the understanding of the structure of DNA and the ability to manipulate it. The idea itself is transcendent; its practical implications are vast. Disease-resistant crops, specially designed drugs, scientific testing procedures, even treatments for hereditary illnesses have now become possible as a result of these technologies. One of the most ambitious projects of the 20th century was to map the entire human genome—to determine the genetic code of DNA in man. Within the span of just a few years, maps of large numbers of individuals were within reach.

In 1951, a young American, James Watson (born 1928), joined Francis Crick (1916–2004) at the Molecular Biology Laboratory at Cambridge University. Both men were motivated by the growing evidence that DNA was the material of genes, and they were determined to discover how its structure would enable it to pass on information from one generation to the next. It was already known that DNA was made up of long chains of alternating sugar and phosphate groups with further chemical groups ("bases") attached to each sugar, but the question was: how were these long chains arranged in space?

The chief experimental evidence was emerging from research being done at King's College London. Maurice Wilkins (1916–2004) and Rosalind Franklin (1920–58) were using the by then well-established technique of x-ray crystallography to study DNA's structure. As with analysis of other such patterns using this technique, the pair was able to examine the fibrous structure of DNA, which gave characteristic x-ray diffraction pictures that indicated the presence of regular arrays of atoms.

Crick had previously mathematically demonstrated that diffraction patterns of a helix (a corkscrew shape) took the form of a cross. The diffraction patterns obtained from DNA fibers by Franklin showed just such a cross shape, convincing Crick and Watson that DNA was helical. It was not, however, possible for them to obtain the complete structure directly from the x-ray data that had so far been obtained, so Crick and Watson turned to model building, in the hope that this intuitive approach might produce a structure compatible with both the x-ray data and the known chemical structure.

In their haste to be the first to solve the problem of structure, Crick and Watson produced, at the end of 1951, a model that included three helical molecules of DNA. However, after Wilkins and Franklin had examined this model, they proceeded to demolish Crick and Watson's arguments for it. In 1952, at the California Institute of Technology, Linus Pauling (1901–94) also turned his attention to DNA, but models that he produced were also shown to be wrong—so the race was still on. Also in 1952, Franklin produced her superb pictures of a second form of DNA. From these pictures, it could be deduced that the structure must contain two chains. It was Crick who then realized that Franklin's analysis indicated that the two chains ran in opposite directions.

The final breakthrough came with chemical evidence provided by Erwin Chargaff (1905–2002) at Columbia University. Four different chemical groups, or bases, could be attached to the sugar groups of the backbone; Chargaff had discovered that the amounts of two bases—adenine and thymine—were always identical. He also discovered that the amounts of the other two bases—guanine and cytosine—were also equal. Watson realized that, as a result of their shape and chemical composition, these two pairs of bases could be arranged with very weak bonds holding them together (the so called hydrogen bonds). The shape of each pair was almost identical, and they would fit down the center of the double helix formed by the two backbone chains.

The elegant structure Crick and Watson proposed offered an immediate explanation of how genetic information contained in DNA could be copied and passed from one generation to the next. The weak hydrogen bonds between each pair of bases, which held the two strands of the double helix together, were easily broken. Each strand then acted as a template for building new molecules. The specific pairing of the bases ensured that each new chain was an exact replica of the original. Crick concluded his letter to *Nature* with: "It has not escaped our notice that the specific pairing we have postulated immediately suggests a possible copying mechanism for the genetic material."

Crick and Watson demonstrated their theories with a series of models. The best known was a large model first shown in 1953. Although the model was dismantled, it was eventually reassembled using its surviving plates (representing the bases) and using new components where necessary.

In 1962, Watson, Crick and Wilkins received the Nobel Prize in Physiology or Medicine. Tragically, Franklin, whose beautiful x-ray diffraction pictures had provided key evidence, had died of cancer nearly five years earlier. By this time, Crick had already started on his seminal work in unraveling the genetic code, enabling the understanding of how DNA specifies the structure of the proteins it is programmed to make. To Crick, as to many other scientists, the structure of DNA explains the emergence and evolution of life. Their model remains a powerful visual argument for that belief. Its significance is expressed in the claim made at the end of the 18th century by the German poet Goethe: "The intrinsic, unique and fundamental theme of world and human history, to which all others are subordinate, remains the conflict between faith and doubt." At the beginning of the 21st century, many surely still agree with Goethe, and in that conflict, Crick and Watson's DNA model is still a vital symbol.

—ANN NEWMARK

Opposite: Reconstruction of Crick and Watson's model made using the original metal plates

FLYING BEDSTEAD TEST RIG (1954)

The rapid developments in aviation technology during World War II led to increases in power, speed and weight for aircraft. Correspondingly, airfields became larger and runways became longer—making them more obvious targets for enemy air strikes. The development of guided missiles after the war made such permanent installations even more vulnerable.

These factors suggested to designers that an aircraft that could rise vertically—or from a very short takeoff run—would have great advantages. Helicopters, of course, could do this, but in the late 1940's they were primitive devices: slow, unable to carry a significant payload, and unsuitable for many military roles. The war, however, had precipitated the operational debut of the gas turbine, and by the 1950's, the new jet engines were showing a marked increase in power output as well as improved reliability. The turbojet could produce a powerful airstream, and it seemed to some aircraft designers that by directing the jet downward, a practical vertical takeoff and landing (VTOL) aircraft might be created.

There was initially little official interest in VTOL aircraft in Britain, but to Rolls-Royce, eager to exploit its new turbojets, the commercial attractions of VTOL were obvious. Rolls-Royce's chief scientist, Dr. Alan Arnold Griffith (1893–1963), was one of the British pioneers of the gas turbine, which he had helped to develop while he was working at the Royal Aircraft Establishment in Farnborough. He had envisioned the use of jet thrust for VTOL aircraft as early as 1941.

Rolls-Royce decided to build two test rigs; they were intended to demonstrate the principles of jet lift to doubters and also to allow for the development of the special control and stabilization systems thought necessary for this new type of aircraft.

The "Flying Bedstead" during tests in Hucknall, Nottinghamshire, c. 1955

Although officially called the "Thrust Measuring Rig" (TMR), the somewhat outlandish appearance of the vehicle led to its being christened the "Flying Bedstead," and that nickname has stuck.

The rig was a tubular framework supporting two Rolls-Royce Nene engines of 5,000 pounds nominal thrust, each with its jet nozzles pointing downward. A seat on top of the engines provided a lofty perch for the pilot. To provide control, the TMR had tubular outriggers that carried compressed air to small supplementary nozzles ("puffer jets") that could be operated by the pilot. An auto-stabilization system was also fitted to the rig, because the aircraft's engineers thought at the time that the pilot's reactions would be too slow to be effective. The total available thrust was 8,100 pounds, an amount that provided little margin for error in a machine weighing 7,500 pounds when it was carrying both pilot and fuel.

The first TMR was rolled out at Rolls-Royce's airfield at Hucknall, near Nottingham, on July 3,

1953, and the first hovering trials followed a week later, with the rig suspended below a specially built gantry. Tethered tests continued over the next year until the first free flight was made by Rolls-Royce's chief test pilot, R.T. Shepherd, on August 3, 1954. Within a few months, the TMR had made sixteen free flights, R.A. Harvey sharing the pilot's duties with Shepherd. The final flight at Hucknall was on December 15, 1954, after which the TMR was overhauled and transferred to the Royal Aircraft Establishment at Farnborough. In the meantime, the second TMR, XK426, had been built. It began tethered hovers at Hucknall on October 17, 1955, and made its first free hover in November 1956.

Both TMRs were damaged in accidents in the autumn of 1957; the second of these accidents killed a trainee pilot. Despite the accidents, the Flying Bedstead program demonstrated, in 380 tethered and 120 free flights, that jet lift could be used to support a fixed-wing aircraft during the takeoff and landing phases of flight, and that puffer jets provided adequate control during hovering.

Rolls-Royce felt confident enough to produce its first engine intended specifically for jet lift, the RB108, installed in Britain's first VTOL airplane, the Short SC.1, which had one engine for propulsion and separate vertical units for lift. Subsequently, the development of the VTOL aircraft took a slightly different path as a result of the creation of the Harrier, in which swiveling nozzles were used to vector the thrust of a single main engine for both lift and forward flight. It was, however, the Flying Bedstead that first pointed the way to successful jet-borne vertical flight.

—Graham Mottram

THE CESIUM ATOMIC CLOCK (1955)

To measure a quantity, one needs a unit. Ideally, this unit should be defined in terms of some naturally occurring phenomenon having universal rather than local significance. For the measurement of time, such a natural unit exists in the length of the day, but, as this unit is too large for many purposes, it has to be subdivided into hours, minutes and seconds to be workable. The length of the day depends on the rate of rotation of the earth on its axis. Astronomical observations have shown that this rotation is not entirely uniform.

During the 1950's, atomic clocks were produced, the timekeeping of which depended on a natural phenomenon, the vibration of cesium atoms. The frequency of these vibrations is so stable that in 1967, the General Conference on Weights and Measures redefined the second in terms of it, thus providing a constant unit of time for specific purposes. Human life is still organized in terms of the day, however, and for civil purposes, atomic time is kept in step with the rotation of the earth by adding or subtracting a leap second, in the same way that an extra day is introduced in a leap year to keep the calendar in step with the seasons.

A cesium atomic clock essentially consists of a tube from which the air has been removed, so that cesium atoms can pass through it. As they do so, they are exposed to radio waves of a very high frequency (comparable to those that are used in radar). Each cesium atom acts as a small magnet, and when the frequency of the radio wave corresponds to the natural frequency of vibration of the atom, the atom flips over, reversing its polarity. A magnetic field deflects these atoms into a detector. By altering the frequency of the radio waves to maximize the number of atoms entering the detector, it is possible to lock this frequency to that of the cesium atom. The frequency of the radio waves is derived, by multiplication, from a quartz oscillator that is thus controlled by the vibrations of the cesium atoms. The quartz oscillator can then be used to indicate the time, as it does in a domestic quartz clock.

The idea of an atomic clock was first put forth by Professor I.I. Rabi (1898–1988) in his Richtmyer Memorial Lecture to the American Association of Physics Teachers in 1945, and most of the pioneering work on the atomic clock was carried out in the United States. In 1948, Professor Polykarp Kusch (1911–93), who worked with Professor Rabi at Columbia University in New York City, produced a design for a cesium atomic clock that was subsequently built at the National Bureau of Standards (NBS) in Washington, D.C. This instrument was operating by 1951, but it never realized the full potential of its design: it was not sufficiently accurate to serve as a time standard. In 1953, Dr. Louis Essen (1908–97) of the National Physical Laboratory (NPL) in England returned from a visit to the NBS and other laboratories in the United States, convinced that a successful cesium clock, based on Kusch's design, could be built at the NPL facility in Teddington, in Greater London. Although the NPL lacked experience in atomic beam techniques, Dr. Essen had made important contributions to other fields (such as narrow bandwidth oscillators), which were crucial to the development of the cesium clock. Assisted by J.V.L. Parry, he commenced work in the spring of 1953, and by June 1955 he had developed a clock reliable enough to have an accuracy equivalent to one second in 300 years. That is more accurate than astronomi-

The 1955 cesium atomic clock

cal observations, which could therefore be supplanted as a standard. To establish a unit of time that was as close as possible to the existing unit, the frequency of the cesium vibrations had to be measured in terms of the best value for the second that could be determined by astronomical observations. He carried out this work between 1955 and 1958 in collaboration with Dr. William Markowitz (1907–98) and Dr. R. Glenn Hall (1921–2004) of the United States Naval Observatory in Washington, D.C. Their figure of 9,192,631,770 vibrations in one second was later used to define the second.

It is reasonable to inquire why the NBS, with its vast resources and expertise, did not succeed in producing a reliable and accurate atomic clock. Paul Forman, of the National Museum of American History, who has made an extensive study of this topic, suggests that the failure was partly due to the division of resources between two different atomic clock projects (an ammonia clock and a cesium clock), which were being developed concurrently. Work on the cesium clock was also disrupted when the laboratory moved from Washington, D.C. to Boulder, Colorado in 1954.

Strictly speaking, the NPL instrument was a time or frequency standard and not a clock, since it did not run continuously or show elapsed time. It is convenient to refer to it as a clock, however, because it operated on the same principle as later atomic clocks, which did run continuously and could show the time. Atomic clocks, as well as technology that relies heavily on them, have certainly become widespread. Portable atomic clocks are available commercially, and are reliable and inexpensive. In addition, atomic clocks make it possible for global positioning system (GPS), radio time signal transmission and Internet application synchronization to function properly—among other systems and devices. Without atomic clocks, it is safe to say that today's technological world—awesome, but often much taken for granted—would be a vastly different one.

—Denys Vaughan

THE WORLD'S FIRST HOVERCRAFT (1955–59)

During the 19th century, scientists recognized that a large proportion of the energy required to drive a ship through water was employed in overcoming the frictional effects of the water on the surface of the hull. Numerous experiments were conducted to reduce this friction, including pumping air over the surface of the hull to create air lubrication and attempting to capture an air bubble upon which the vessel could rest. All these experiments proved unsuccessful.

In 1950, C.S. (later Sir Christopher) Cockerell (1910–99) left his employment as a research engineer with the Marconi Company and acquired a small shipyard on the Norfolk Broads near Lowestoft, Suffolk, England. His background as an engineer led him to investigate the high frictional resistance of boats, and, after a number of experiments with air lubrication, he came to the conclusion that the best method of tackling the problem was to "float" the hull on a cushion of low-pressure air. Unlike previous attempts, which had failed because of too much air leaking from the cushion when the craft heeled, Cockerell's method both contained the cushion and kept the craft stable. His breakthrough was to create a "curtain" of high-pressure air ejected downward from the rim of the craft.

In 1955, to test this idea, Cockerell built a working model that used a model-aircraft engine to provide both the supporting cushion of air and the containment curtain; the engine also provided forward thrust. The model worked well, and Cockerell formed a company to develop the idea.

In 1956, the Research Branch of the British Ministry of Supply took an interest in the project and signed a contract with Saunders-Roe Ltd., the makers of many famous flying boats, to undertake a study of the concept in conjunction with Cockerell. This study was classified as "secret," and had as its main purpose the identification of the invention's military potential. The results indicated the viability of the concept, although no sufficiently important military role was identified—at least none that would warrant further military funding. In 1958, the project was declassified, and the National Research Development Corporation (NRDC) decided to finance further development work through a new company called Hovercraft Development Ltd. In the autumn of 1958, the NRDC signed a contract with Saunders-Roe to create the world's first full-size hovercraft with the designation SR-N1.

The SR-N1 was, from the beginning, conceived as an experimental craft, and upon its completion in June 1959 it consisted of an oval platform thirty-one feet in length with a breadth of twenty-five feet. It hovered at a height of about one foot, thus earning itself the nickname "the flying saucer." As in the model, the cushion, curtain and forward thrust were provided by a single-piston, 435-horsepower

Top: Testing the model shown opposite.
Above: SR-N1 hovercraft, Calais, July 24, 1959

Alvis Leonides engine. The craft carried a pilot and an observer, and it demonstrated its ability to operate over both land and sea.

Although the SR-N1 crossed the English Channel from Calais to Dover in July 1959, it became clear as the experimental program progressed that a hovering height of one foot restricted it to virtually smooth ground on land and to a two-foot wave height at sea. In addition to these problems, the shape of the platform led to difficulties in steering the craft. These directional problems were gradually solved by the installation of a separate jet engine to provide forward thrust and by the addition of a pointed bow and stern and an inflated keel. The problem of height clearance was solved by the development of the "flexible skirt," a rubberized fabric that hung from the platform rim, through which the curtain jets were led. By means of the skirt, the operational height of the craft was raised from one foot to four feet.

The SR-N1 completed its experimental program in early 1964, and from those experiments a series of hovercraft were developed that were capable of operating over both land and sea. Perhaps the best known of these were the SR-N4 (Mountbatten class) cross-channel passenger ferries that entered service in 1968 and carried 250 passengers and thirty cars. These craft, in various extended versions, operated on this route until 1991.

Unfortunately, the early promise held by the hovercraft as both a mass-transport vehicle and a new industry was not fulfilled in Britain, principally because of the high costs of manufacture, maintenance and fuel—all legacies of the craft's aircraft origins. Although attempts were made in the 1980's to remedy these shortcomings, the hovercraft of the kind represented by the SR-N1 has only been able to operate in specialized world markets. Most of its uses today are for search-and-rescue, military and recreational purposes.

—Tom Wright

AMPEX VR-1000 VIDEO RECORDER (1958)

Magnetic tape recording, as distinct from previous systems using steel tape or wire (*see* **Poulsen's Telegraphone**), was developed in Germany during World War II. Magnetophon tape recorders captured from the Germans, along with the instructions for the manufacture of the tape, formed the basis for all immediate postwar research.

The development at this time of instrumentation recorders capable of recording vast amounts of data encouraged the belief that video recording was a technical possibility. A number of research establishments throughout the world were engaged in work toward this end.

By 1950, the Ampex Company in California was well established as a manufacturer of broadcast audiotape recorders, and the company was developing a multi-track instrumentation recorder. At the start of 1952, a project to investigate video recording had been set up under Charles Ginsburg (1920–92). He was soon joined by a young engineering student, Ray Dolby (born 1933), whose name was later to become synonymous with noise-reduction systems. By the autumn of that year, the two were able to demonstrate a device that could record almost recognizable pictures. The project was then abandoned for more than a year, but in late 1954, Ampex redoubled its efforts with the creation of a larger team that was able to solve the remaining problems of the video recorder within eighteen months.

One of the many obstacles that had to be overcome was the problem of tape speed. In order to record the vast quantity of information in a television picture, the tape had to pass the recording head at a very high speed, and it was this problem that caused most other designs to founder. In 1953, RCA demonstrated an experimental machine that ran at 360 inches per second and rushed one and a half miles of tape through the recorder—but provided just four minutes of playing time. The team at Ampex used a rotating-head technique to solve the problem. Four heads, mounted on a rotating wheel spinning at high speed, scanned the slowly moving tape. Although the tape speed was only fifteen inches per second, the relative head-to-tape speed

*Ampex videotape recorder,
type VR-1000A, c. late 1950's*

was 1,600 inches per second, which provided for one hour of recording on a ten-inch spool.

Ampex had kept this technological breakthrough a secret, so when they gave a surprise demonstration in April 1956 at the convention of the National Association of Broadcasters in Chicago, the industry was taken by storm. Ampex took orders for eighty machines within the first four days after the demonstration—advance orders worth $4 million.

The first broadcast use of the machine was by CBS on November 30, 1956. By early 1958, a version of the machine was available in the U.K., and machines were purchased by Associated-Rediffusion and Granada Television, with the BBC following suit the year after. The Ampex system was so successful that it rapidly killed off the competition to the extent that several companies (notably RCA) signed licensing agreements to allow them to market machines using the same format. The Ampex format was to reign supreme in broadcast television for more than twenty years, and it fundamentally changed the way in which television companies operated.

In the mid-1950's, the vast majority of television programs were broadcast live, a method that

involved complex extended rehearsals, production crews and presenters working shifts, and interludes to allow resetting of studios between programs. Most importantly, program choices and styles were restricted to formats that would suit live television. From quiz shows to sophisticated drama, television was ephemeral.

Program planners also had to cope with externally scheduled events that had to be inserted into the program schedule. For example, with a sporting event, altering the broadcast time was impossible without video recording. The headache of coping with a boxing match that might end halfway through the first round was equaled by the problem of what to do when a game overran its time slot. Every television station had its stack of emergency caption cards, which attempted to compensate for the unexpected.

The move away from live television toward pre-recording took many years, largely due to the high cost of video-recording technology shouldered by broadcasters in the 1960's and 1970's. However, this gestation period for the industry's video recorder was a precursor to the much larger revolution in television that occurred following the introduction of the video cassette recorder (VCR) for the home in the 1980's. The VCR was responsible for a wholesale shift in television viewing habits. Of course, with the subsequent introduction of the DVD recorder, and then the digital video recorder (DVR), not to mention access to reruns on cable television and on the Internet, the days when it presented a problem when a program's scheduled time clashed with a viewer's own schedule, or clashed with another desired program on another channel, became but distant memories.

The early Ampex video recorder is a milestone in the history of television for many reasons, not the least of which is the enormity of what it spawned. It marked the beginning of the end of the prevalence of live television, but it heralded so much more for the future realm of television, culture and society.

—JOHN TRENOUTH

FERRANTI PEGASUS COMPUTER (1959)

The Ferranti Pegasus computer was an early computer that predated the use of transistors or integrated circuits. It is a tribute to the engineering design skills at Ferranti Ltd. that the machine pictured opposite is still fully functional, although it was built in 1959.

The Pegasus is most impressive to see and to hear when it is running. Its valve circuitry consumes eighteen kilowatts of power, provided by a loud generator, while air-conditioning units roar to keep it cool—all to achieve the computing power of a modern handheld calculator. But when it first appeared in 1956, it was capable of performing calculations infinitely faster than could adding machines powered by human operators—the only alternative form of calculation generally available at that time.

The Pegasus computer used what was, for the time, a novel form of construction in which the circuitry was built from a number of modular packages. The package concept was first developed by the firm of Elliott Brothers in Borehamwood, Hertfordshire, England and came about as a result of a Royal Navy requirement for an electronic gun-control computer. Previous computer designs had involved assemblies of components wired together as one or two large units. For ship-borne operations, however, it was not possible to carry spares of such large units; thus, in the Elliott system, the circuitry was built up from a large number of packages of which there were only a small number of different types. Each package within a type was interchangeable, therefore only a small number of spares had to be carried: a faulty package could be replaced and then subsequently repaired somewhere other than on board a ship.

The package concept had many critics, however. Some said that the plugs and sockets needed at the rear of packages could be unreliable; others said that the use of packages would restrict the logical design of a computer. All of these criticisms

Ferranti Pegasus computer, 1956

proved to be unfounded, as Pegasus so successfully demonstrated.

The designer Christopher Strachey (1916–75) was at this time working at the National Research Development Corporation (NRDC), and he was an enthusiastic supporter of the package concept. Basing his own design on experimental work he had done with Elliott Brothers, he proposed the basic architecture for the Pegasus.

Strachey also proposed a new kind of logic design, the "order code" for the computer. This was a major advance in clarity and simplicity. The order code is the way in which a computer is instructed to carry out the very limited actions that it can perform. The computer uses binary information: ones or zeros. Humans cannot easily express themselves in this way: it is the order code that mediates between human intentions and the computer's internal processes.

In devising his logical structure for the machine, Strachey was following the example of Maurice Wilkes (1913–2010), one of the leading British computer pioneers, who designed and built the Electronic Delay Storage Automatic Calculator (EDSAC) computer at Cambridge University in the late 1940's. Wilkes stressed the importance of the order code; he placed great emphasis on ease of programming even at the expense of computing speed. As a result, Strachey made sure that Pegasus was a machine that was easy to program and logically very "clean;" it was much loved by all who worked with it. Strachey's order-code structure continued in the late 1970's with the ICL 1900 range of mainframe computers.

The first Pegasus machine, the Pegasus I, was assembled in very elegant surroundings in Portland Place in London in 1956. It proved very reliable, compared with other computers of the time, and it was soon providing the first bureau service in Britain, selling computer time to outside customers.

The Pegasus I evolved into the Pegasus II, which, although it involved the same outward appearance, featured improved circuitry, a larger drum store and more peripherals, including a printer and magnetic-tape units. In all, thirty-eight Pegasus units were sold, many going abroad. The Pegasus was one of the first computers to be produced on a production line, set up in Ferranti's factory at West Gorton in Manchester.

The Pegasus pictured opposite (number 18) was first sold to Scania in Sweden in 1959. In 1963, it came back to Ferranti, the computer division of which had by then become part of Standard Telephones and Cables Ltd. It then went to the Chemistry Department of University College London, where Dr. Judith Milledge used it until 1983 for the analysis of x-ray crystallography results. In 1983, it was acquired by the Science Museum, London, where, with the assistance of former Pegasus maintenance engineers, it was restored to full working order by the summer of 1990.

—Tony Sale

ELECTRON-CAPTURE DETECTOR (LATE 1950'S)

While the media spotlight was on the "big science" of atom-smashers and space missions, much of British science in the 1950's was a matter of making gadgets out of sealing wax, string and recycled machine parts. But no "homemade" instrument of the 1950's has had a bigger impact on the world than the electron-capture detector (ECD).

For decades, chemists had been struggling to separate the complex mixtures generated by the petrochemical industry at one end of the scale and the biochemistry of cells at the other. At the same time, they were also attempting to detect chemicals at very low concentrations, at parts per million or even parts per billion. Existing methods were incapable of achieving these goals. The separation problem was solved by the development of gas-liquid chromatography by Archer Martin and Tony James at the National Institute for Medical Research (NIMR) at Mill Hill, a northern suburb of London, in the early 1950's. This powerful technique sorted out the components of a gaseous mixture by passing it through a glass tube coated with silica. But this breakthrough created another problem: there was no easy way of detecting the different compounds as they came out of the chromatographic apparatus.

A young researcher at the NIMR, James Lovelock (born 1919), encountered this problem when he tried to use gas chromatography to separate a mixture of fatty acids extracted from cells. He sought the advice of Martin and James, who—half in jest and half in seriousness—told him to create his own detector. Lovelock quickly developed a detector that was very similar in principle to a modern smoke detector: an electrical potential is created across a narrow gap between two electrodes. Lovelock's detector also contained a radioactive source that knocked electrons off certain molecules as they entered the detection chamber and subsequently produced charged molecules, called ions. These ions allowed a current to pass across the gap, which was then detected and amplified. Initially, Lovelock used nitrogen gas to sweep the compounds from the chromatograph into the detector, but when at one point he was forced to use argon instead, he found that argon as the carrier gas greatly improved the performance of the device. This enabled Lovelock to create the argon detector—which was the forerunner of the ECD. In the 1950's, working with the American medical researcher Seymour Lipsky (1924–86), Lovelock developed the modern ECD, which was quickly commercialized.

When Lovelock had attempted to demonstrate the early version of his detector to Martin and James, he had been discouraged by the large number of "hits" (called peaks) generated by a mixture of a few chemicals. By the time he had developed the ECD, he realized that these false peaks were produced by tiny amounts of organic compounds that contained chlorine, which were easily converted into ions by a radioactive source. When large numbers of fish were dying in the Mississippi River in 1963, scientists, using standard chemical methods, were unable to discover the toxic agent. The then novel ECD quickly showed that the fish had been killed by a mixture of chlorinated hydrocarbons, which were then traced back to a pesticide factory in Memphis, Tennessee. Within a decade, the ECD was detecting traces of pesticides at the level of parts per billion—the equivalent of half a teaspoon of DDT in an Olympic-sized swimming pool.

Meanwhile, Lovelock had taken an ECD to his holiday home on the west coast of Ireland. To his astonishment, he was able to detect chemicals called CFCs that had traveled across Europe. CFCs (chlorofluorocarbons) had been developed for use in refrigerators and aerosols in the 1930's and were chemically inert. When the wind was from the west, he detected a lower level of CFCs and wondered if they had come from North America—or if the atmosphere generally contained a stable level of CFCs. Given their chemical inertness, Lovelock suspected that the latter explanation was true, and he sailed to the relatively unpopulated South Atlantic to find out. He discovered that there was indeed a stable level of CFCs in the lower atmosphere, and his results were used by F. Sherwood Rowland (1927–2012) and Mario Molina (born 1943) in their historic paper on ozone depletion published in the British journal *Nature* in the June 28, 1974 issue.

Rowland and Molina claimed that CFCs, though usually inert, reacted with ozone in the upper atmosphere in the presence of sunlight, and that these CFCs consequently reduced the thickness of the ozone layer—which protects life on earth from the sun's harmful ultraviolet rays. The manufacturers of CFCs attacked their findings, but these same manufacturers were forced to back down and phase out the production of CFCs when the British Antarctic Survey showed in 1985 that the ozone layer above the South Pole had indeed become dangerously thin. Thanks in part to Lovelock's homemade electron-capture detector, the problem of ozone depletion was on its way to being solved—before it could cause an ecological catastrophe.

—Peter Morris

INTERNAL CARDIAC PACEMAKER (1961)

The development of the internal cardiac pacemaker is a tale of combined effort rather than individual endeavor. Several groups were working in the same area of clinical research at the same time, and all of them sought to solve, in different ways, the problem of artificial pacing of the heart. The model pictured opposite is an early British contribution to the field.

Researchers had been active in this field for several years following the pioneering work of Paul Zoll (1911–99) in the United States. In 1952, Zoll was the first person to resuscitate a human heart using an electrical stimulus. In 1958, the Swedes Rune Elmqvist and Åke Senning, working in Stockholm, developed and implanted the first pacemaker in a human. Their unit was not completely internal: it relied on an external power source to recharge the internal system. The first completely self-contained internal cardiac pacemaker was reported in October 1960 by a team from Buffalo, New York, comprising William Chardack, Andrew Gage and Wilson Greatbatch. In London, a team based at St. George's Hospital, including doctors Aubrey Leatham and R.W. Portal and surgeon Harold Siddons, were treating a number of patients with Stokes-Adams disease, or heart block. Other research groups were looking at a variety of solutions, including using radio pulses or radioactive isotopes to produce a regular stimulus for the heart.

A complete cycle of the contractions of the heart is comprised of two separate stages: first, the auricles (the upper chambers of the heart) contract, pushing blood down into the ventricles (the lower chambers); next, a second contraction pushes blood out of the ventricles into the arteries. In a normal heart, the natural cardiac pacemaker generates an electrical pulse that initiates each cycle. This pulse is transmitted down through the wall dividing the two sides of the heart and on to the ventricles, leading to their contraction.

In a patient with heart block, the system conducting the pulse is impaired, and this condition can lead to the heart's temporarily stopping. When this stoppage occurs, the patient has had a Stokes-Adams attack. Typically, the person suddenly loses consciousness, collapses and has a seizure. The similarity of the symptoms meant that Stokes-Adams disease was occasionally confused with epilepsy.

In order to regulate the contractions of the heart artificially, an electrode needed to be inserted so that its tip was touching the inside of the right ventricle. This procedure was accomplished by cutting into the jugular vein and threading the end of the wire electrode down through the blood vessels and into the heart. A temporary external pacemaker was than connected to the other end of the wire, which remained outside the body. To complete the circuit, a second electrode needed to be attached to the outside of the chest.

At St. George's Hospital, the team initially experimented with small external pacemakers that could be carried around by the patient, but the problems with infection caused by keeping a permanent opening in the body for the wires led

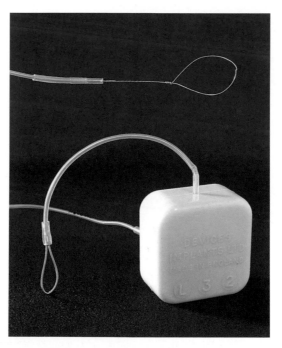

St. George's Hospital's implantable axilla cardiac pacemaker, 1967

the team to try to devise internal models. The first design consisted of an implanted receiver and an external radio transmitter set into a belt that the patient wore around the waist. This design was only partially successful, as it still required the patient to carry an apparatus, and a powerful radio pulse was needed—otherwise, the internal receiver couldn't pick up the signal. Geoffrey Davies, a senior cardiac technician at St. George's, then designed the completely internal pacemaker that became the main device used at the hospital.

The device was powered by mercury-zinc batteries that had an estimated life of five years. In clinical use, the entire unit was intended to be replaced every eighteen months. The pacemaker was implanted under the skin and two electrodes were attached to the heart during a major operation in which the chest wall was opened. Later, it became possible to insert electrodes for long-term pacing. To accomplish this, an electrode was threaded through the jugular vein—as was done for temporary pacing. However, the pacemaker and the second electrode were then implanted under the skin, and the electrode in the vein was connected to the heart in what became a much easier and safer procedure.

At first, because of the relative rarity of heart block, physicians thought that the demand for the internal cardiac pacemaker would be low. However, later developments in the technology meant that use of the device was not only possible, but crucial so that some patients could survive a variety of other cardiac conditions in which control of the heart's rhythm was affected. Today, it is estimated that more than three million people around the world have an implanted pacemaker.

—JANET CARDING

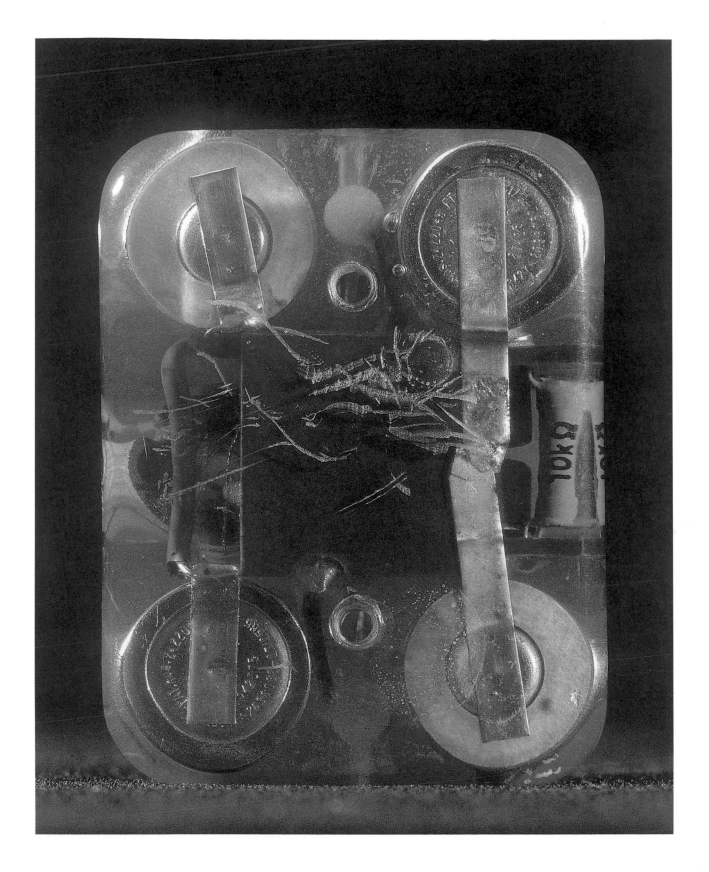

SCANNING ELECTRON MICROSCOPE (1965)

Intuitively, one might think it possible to see any object, however small, simply by enlarging its image under a microscope. But microscopes cannot give a clear image of objects that are much smaller than the wavelength of the illumination that is used to form the image. That means, in effect, that we cannot see objects that are smaller than approximately one ten-thousandth of a millimeter under a light microscope. To see smaller objects, we must use a source of illumination that not only has a shorter wavelength, but also can be focused by a lens. Electrons satisfy both of these requirements—the wavelength of a fast-moving electron is several orders of magnitude shorter than the wavelength of light, and it can also be deflected by an electromagnet, thus reproducing the effects of a lens.

The first electron microscope was built by Max Knoll (1897–1969) and Ernst Ruska (1906–88) in 1931; it evolved from a research project on the high-voltage oscillograph at the Technische Hochschule in Berlin. This instrument operated in the same way as a light microscope: the electrons passed through the specimen, and an enlarged image was formed by the electromagnetic lenses. The scanning electron microscope, however, builds up the image, bit by bit over a period of time. This result is achieved by scanning a fine beam of electrons, with a zigzag motion, across the surface of the specimen, then collecting the electrons deflected from the surface of the specimen by the electron beam. These electrons are used to control the brightness of a spot of light—on a video screen—that moves in tandem with the beam. The scanning electron microscope therefore has the advantage of being able to examine the surface of specimens directly. The size of the electron beam determines the fineness of the detail, and the magnification can be altered by varying the area of the specimen that is being scanned: the smaller the area, the higher the magnification.

Scanning electron microscopes, operating on these principles, were constructed by Manfred von Ardenne (1907–97) and Max Knoll during 1933

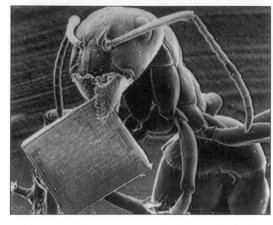

A scanning electron micrograph, showing the depth of field in such images (Philips)

and 1934. In the following years, von Ardenne attempted to see finer detail by using an electromagnetic lens to reduce the size of the electron beam, but the quality of the image was poor. In 1948, Charles (later Sir Charles) Oatley (1904–96) felt it timely to reconsider the feasibility of the scanning electron microscope in light of the advances in electronics that had occurred during World War II and shortly afterward. He initiated a research project at the Engineering Department of Cambridge University in England, and by 1951, his research student Dennis McMullan had produced a working instrument with a resolving power that exceeded that of the light microscope. This was the forerunner of the modern scanning electron microscope. The work on this instrument was then taken over by another research student, K.C.A. Smith, who dramatically improved the efficiency of the electron-collecting system.

The Pulp and Paper Research Institute of Canada was impressed by the results obtained with the improved microscope, and it employed Smith to construct a properly engineered version, which worked successfully. Despite the success of the

Canadian instrument, however, a further seven years were to elapse before the instrument was produced in quantity. Initially, there simply wasn't a perception that there would be a sufficiently large market for it, since the conventional electron microscope, which by then was firmly established, could see finer detail and examine surfaces indirectly. As for viewing specimens at lower magnifications, these needs were taken care of by the ubiquitous light microscope.

Eventually, the Cambridge Scientific Instrument Company agreed to manufacture the scanning electron microscope; the company was influenced by the fact that some of the parts needed to do so were already available in another instrument that they were producing. One of Oatley's research students, A.D.G. Stewart, joined the company to work on the microscope, thus maintaining the link with the university. In 1965, in order to test the market, the company decided to produce a trial batch of five instruments. The fourth, pictured opposite, was originally used by the Central Electricity Research Laboratories.

The five were sold under the brand name "Stereoscan." This name was chosen to emphasize the strikingly three-dimensional quality of the images, which were enhanced by the tremendous depth of field—far greater than that of the light microscope at the same magnification. It was only when these instruments came into the hands of microscopists that this feature (and others, such as the ease of specimen preparation) was fully appreciated. Sales boomed, and similar instruments were soon produced by other companies. By 1985, annual production had risen to a thousand instruments—a far cry from the ten instruments that a 1963 American market survey predicted would be needed annually.

—Denys Vaughan

ULTRASOUND SCANNER (1967)

It is now an eminently routine practice for pregnant women to have ultrasound scans. Yet the first experimental ultrasound machine for use in gynecology and obstetrics was constructed as recently as 1956. In the years since, what was once a novel and experimental procedure has become part of the normal experience of pregnancy. The Diasonagraph Mk1, from the Clinical Institute of Obstetrics and Gynaecology, Imperial College London, played a key part in this revolution.

Ultrasound is sound higher in pitch than can be heard by the human ear. An ultrasound scanner uses crystals that naturally resonate at these high frequencies to transmit and pick up sound waves. In the first experiments with ultrasound, undertaken during World War I, the French physicist Paul Langevin (1872–1946) aimed to develop a technique for detecting submerged German submarines. Largely unfinished by the end of the war in 1918, the work continued, with important developments occurring during the intensification of submarine warfare in World War II. In 1928, ultrasound theory was also applied to detecting flaws in metal components.

The first published experimental ultrasound examinations of women, undertaken by Professor Ian Donald (1910–87) and his colleagues in Glasgow in the late 1950's, used an industrial flaw detector to show that echoes from within a patient's abdomen could be used to measure the size of ovarian cysts and other tumors. This type of scan is known as a one-dimensional scan because it shows how far apart objects are without actually producing an image. In two-dimensional scans, actual images are produced by the technique of moving the ultrasonic probe in a series of sweeping movements during scanning. Both techniques are now used when a patient goes to a medical facility for a scan.

When ultrasonic waves are passed from a scanner into the body, they bounce back like echoes when they meet the interfaces between different kinds of tissue. The scanner picks up these reflected sound waves and displays them on a screen. Objects that are nearer the surface of the body send back echoes before those that are further away, so that what appears on the screen corresponds exactly with what is inside the body. Objects that are denser send back stronger echoes, and these differences allow doctors to distinguish, for example, bone from muscle.

Professor Stuart Campbell (born 1936), using the machine that is pictured opposite (built in about 1967) established many of the procedures that have made ultrasound an indispensable part of routine prenatal care. In a period of heightened awareness of the dangers of x-rays, ultrasound was seen as a safe way of examining all pregnant women for multiple fetuses or for deformities. In 1969, quintuplets (later successfully delivered) were diagnosed at nine weeks using the two-dimensional scan. The first abortion of a malformed fetus after an ultrasound examination occurred when the scanner pictured opposite disclosed a case of anencephaly (the partial or total absence of a brain) in 1972. In 1975, ultrasound scanning proved capable of diagnosing spina bifida.

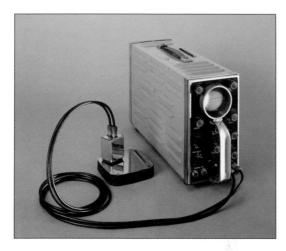

Ultrasonic flaw detector, c. 1959

Although checking for gross abnormalities and multiple fetuses is now an important part of a scan, perhaps more significant is the use of these machines to produce measurements of the fetus that allow comparison with established patterns of growth. During the late 1960's and into the 1970's, Campbell, using the Diasonagraph, developed a number of these techniques. In London, he built upon earlier work he had undertaken with the Glasgow team in which the fetal head was measured by first accurately establishing its position in the womb, using a two-dimensional scan, then measuring it, using the one-dimensional technique. By comparing the size of the head with established norms, Campbell was able to estimate the age of the fetus and to judge whether it was an appropriate size given its age. Other techniques followed, including one that ascertained the fetus' hourly urine production from the volume of the bladder. A low urine-production rate was found to be linked with reduced fetal growth. It was also possible to "weigh" the fetus by measuring its abdominal circumference, once again relating this reading to average growth rates in order to establish whether a fetus could be considered an appropriate size for its age.

The rapid acceptance of ultrasound scanning, along with other new obstetric technologies (notably fetal monitoring during childbirth) was sometimes criticized. Some argued that by inventing a technology that made it possible to "know" the fetus independently of the mother, doctors bypassed the one person who might be expected to know the most about the pregnancy. Studies have indicated, however, that having a scan helps some mothers to bond with their babies. It is also entirely possible that the ultrasound scanner, because of its role in establishing core techniques of prenatal care, has helped patients to accept more readily the part played by such machines in their medical treatment.

—Tim Boon

THE MOLINS SYSTEM 24:
THE FIRST FLEXIBLE MANUFACTURING SYSTEM (1967)

The Molins System 24 was a British manufacturing concept ahead of its time: it combined emerging computer technologies and the numerically controlled machine tool into an integrated system that could manufacture components with very little human intervention. The result was one of the biggest single developments in the history of production technology in the 20th century.

Modern mass-production techniques trace their ancestry back to the small workshops that appeared during the first half of the 19th century, and although the machine tools—the lathes, milling machines, etc.—have become more sophisticated, they are strikingly similar to those designed by such inventors as Whitney, Whitworth and Maudslay. The Portsmouth block-making machinery (*see* **Portsmouth Block-Making Machinery)** was the world's first mass-production line (c. 1803), and this basic concept remained largely unchanged until the middle of the 20th century. The advent of numerical control in the 1950's, whereby machines could be automatically controlled by coded instructions, helped to improve the efficiency of machining, but it did little to streamline the manufacturing process for each component. The introduction of the digital computer in the late 1950's was the integrating element that could replace human control of individual machine tools, thus enabling the manufacturing process to be entirely automatic.

In 1965, a research and development engineer, D.T.N. (Theo) Williamson (1923–92), who worked for the Molins Machine Company, tobacco machine manufacturers in Deptford, southeast London, first integrated these elements into a computer-controlled flow-production system, the System 24. Williamson had been one of the pioneers of numerical control (NC) in the United Kingdom, and he had published the first British paper on the subject in 1955, while working for the electronics firm Ferranti in Edinburgh (*see* **Ferranti Pegasus Computer**).

The first prototype System 24 machine tool began to cut metal in 1964, and a further range of models was developed between 1966 and 1968. A series of patents was granted from about 1970. The concept of System 24, as Williamson envisioned, was that of a number of complementary machine tools ("units") under the control of a small digital computer (which also controlled a conveyor system) so that work pieces could be automatically moved from unit to unit for different machining operations. The idea was for the system to run unattended throughout a sixteen-hour night shift. Maintenance operations were required only during the day, and primarily consisted of loading and unloading the system. The computer controlled all the handling equipment and the processes of work-setting, resetting, inspection, unloading, program-tape selection, cutter changing, and loading and rescheduling in the event of machine failure. Each

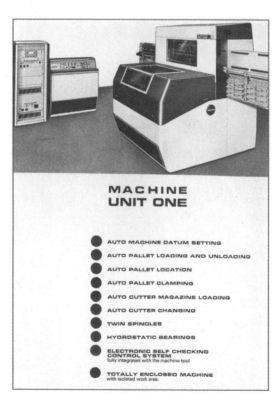

MACHINE UNIT ONE

● AUTO MACHINE DATUM SETTING
● AUTO PALLET LOADING AND UNLOADING
● AUTO PALLET LOCATION
● AUTO PALLET CLAMPING
● AUTO CUTTER MAGAZINE LOADING
● AUTO CUTTER CHANGING
● TWIN SPINDLES
● HYDROSTATIC BEARINGS
● ELECTRONIC SELF CHECKING CONTROL SYSTEM
  fully integrated with the machine tool
● TOTALLY ENCLOSED MACHINE
  with isolated work area.

A page from the Molins System 24 brochure, 1971, showing the Unit 1 machine (Molins PLC)

unit in the line had different functions, and held magazines of between eleven and twenty-eight tools. Taken as a whole, the line was capable of complex metal-cutting operations, all linked by an accurate component transfer and location system. In addition, the system was designed to use binary coding for identifying the pallets that carried the components from one location to another. It was also designed to manufacture parts from aluminum alloy and to achieve as high a productivity rate as possible.

Only one full prototype system was manufactured—for IBM—but components of System 24 were sold to a number of companies, including Texas Instruments and Rolls-Royce. British Aerospace (Wharton Division, Preston) had six Unit 1 machines up until the mid-1980's, and they were arranged in a cell system, similar to the original Williamson concept. Little else of Williamson's original idea was effectively put into practice, however, as technological advances in cheap microelectronics in the United States during the late 1960's and early 1970's enabled companies such as Cincinnati Milacron to develop simpler, less costly manufacturing systems.

In the end, the overall concept proved too expensive for Molins to develop to its full potential, and the company eventually dropped the scheme. The influence of System 24, however, quickly spread throughout the production industry. Remnants of Williamson's ideas can be seen in advanced FMS (Flexible Manufacturing Systems) and in the field of CIM (Computer Integrated Manufacturing). In 1972, Williamson gave a review lecture entitled, "The Anachronistic Factory," in which he anticipated the advent of the automated factory years before it became a reality. Certainly, the developments in automated manufacturing that now exist owe a great deal to the pioneering work and original thinking of Theo Williamson and his team at Molins.

—John Griffiths

APOLLO 10 COMMAND MODULE (1969)

The Apollo 10 command module, call-sign "Charlie Brown," carried three astronauts— Commander Tom Stafford, John Young and Eugene Cernan— into lunar orbit in May 1969. The flight was a final test mission before the planned lunar landing of Apollo 11 in July. It was the first time that the Apollo spacecraft and its systems had been tested for a landing on the moon.

The mission confirmed that "lunar orbit rendezvous"—the technique chosen by the National Aeronautics and Space Administration (NASA) to fulfill President Kennedy's commitment to taking an astronaut down to the lunar surface before returning him safely to earth—would work. This method required two spacecraft—a "lander" (lunar module) and a command module (also attached to a service module) to make the 240,000 mile voyage to the moon. The two craft would separate in lunar orbit, and the lunar module and its crew of two would then descend to the surface of the moon, leaving the command module and its crew of one orbiting above. When the two astronauts had completed their surface exploration, they would blast off in the lunar module's upper half (the ascent stage) to rendezvous with their colleague in the command module.

On the Apollo 10 mission, Stafford and Cernan (the pilot) took their lunar module, call-sign "Snoopy," to within about nine miles of the moon's surface, investigating what was to be the Apollo 11 landing site and taking hundreds of photographs of what Stafford had initially described as a "big gray plaster of Paris thing." When they were finished, they fired their engine to boost them back into orbit and to a reunion with Young and the command module. The separation, the swoop down to the moon, the re-docking—all hinged on the two spacecrafts' respective radar and navigation systems performing to plan; they had never before been tested in lunar space. They performed almost perfectly, and Apollo 10 thus paved the way for the triumph of Apollo 11 two months later.

"Charlie Brown" was a Block II-type command module (the Block I had no docking facility and was capable of flying only in the earth's orbit). The crew compartment was formed from an airtight double shell of aluminum alloy. Wrapped around this shell was an aft and central heat shield (the forward shield was jettisoned soon after reentry into the earth's atmosphere in order to allow deployment of the parachutes).

The heat shield comprised an outer layer of epoxy resin, hand-injected into a fiberglass honeycomb of approximately 400,000 cells and a steel honeycomb inner layer. It was thickest at the rounded aft end of the spacecraft, which struck the atmosphere first (at a still unmatched human velocity record of 24,790 miles per hour), and so experienced the greatest amount of heat as it squashed the air in front of it; the charring is still clearly visible where almost half the thickness of the shield burned away.

Pairs of maneuvering rocket nozzles can also be seen around the outer heat shield. They were used to orient the spacecraft through axes of pitch, roll and yaw, particularly when Stafford lined up the spacecraft for its critical angle of reentry into the atmosphere. In addition, one can now see that the spacecraft's hatch door has been removed, but nearby handles set into it would have been used by the astronauts during any emergency space walks.

"Charlie Brown" now belongs to the National Air and Space Museum of the Smithsonian Institution in Washington, D.C. NASA donated it to the Smithsonian in 1970, and it toured Europe for several years before the Smithsonian loaned it to the Science Museum in London. It is an artifact from the Apollo 10 mission, but it is now venerated as much for its representation of Apollo 11 and its culmination and realization of President Kennedy's dream.

And yet, "Charlie Brown" represents far more than that ultimate triumph; it also represents an almost unimaginably complex assemblage of human activity that was marshaled, coordinated and directed toward the creation of what was then the world's most complex machine, one that would be able to keep three humans alive and in comfort during their journey to and from the moon—and also serve as their mission's control center. Ultimately, the Apollo 10 command module did its job—and its near flawless flight record means that we may not have seen the last of its kind.

—DOUG MILLARD

CONCORDE 002 (1969)

The origins of Concorde can be traced back to the early 1950's. Jet engines were being developed rapidly, and swept and delta (triangular) wing shapes were being explored for supersonic military aircraft. In March 1956, the potential of the new technology was demonstrated when the Fairey Delta 2, powered by a Rolls-Royce Avon turbojet engine, set the world airspeed record at the then astonishing speed of 1,132 miles per hour (Mach 1.7).

The first jet-powered civil airliner, the de Havilland Comet, which entered service in 1952, brought a significant reduction to international flight times. The trend in civil aviation since the end of World War I had been toward achieving faster aircraft speeds, and it seemed a logical step to consider supersonic travel. To this end, the Royal Aircraft Establishment (RAE) in Farnborough, England initiated design studies.

In November 1956, the British government created the Supersonic Transport Aircraft Committee, and in 1959 that committee recommended a major step forward: the construction of a fleet of long-range intercontinental airplanes that would fly nearly twice as high (50–60,000 feet) and more than twice as fast as aircraft in service at the time. The design speed was to be Mach 2, twice the speed of sound.

The French had also been pursuing the idea of a supersonic airliner, and in 1962 the British and French governments signed a binding agreement to develop the aircraft together. Two prototypes, 001 (French) and 002 (British) were built to put theory into practice.

Above all else, Concorde had to be a safe aircraft, and, from a pilot's point of view, it had to handle in a conventional manner at all speeds: flying at the speed of sound (Mach 1) and at twice that speed (the equivalent of about twenty-three miles per minute) had to be no different from flying at subsonic speeds. Every potential problem was considered, and backup systems were provided for every operation that might fail. For example, on Concorde, there were four different ways of lowering the landing gear.

A wind tunnel model of Concorde undergoing tests at the Royal Aircraft Establishment, Bedford, England, in 1962

A maximum speed of just over Mach 2 was chosen, because aluminum alloys could be used. At this speed, Concorde's skin temperature could reach 248 degrees Fahrenheit, the result of air friction. To fly at higher speeds, titanium and stainless steel would have been needed, and they would have involved more costly research and development.

Concorde 002's complex and elegantly curved wing was designed to meet the requirements of low drag at high speed and high lift at low speed. As well, the wing contained eighty percent of the total fuel load of 27,739 U.S. gallons (105,000 liters).

Many innovative technical solutions had to be developed for the program. Concorde's delta wing created a pronounced nose-high attitude at low speed, and to improve the pilot's view when landing and taking off, engineers devised a mechanism to drop the nose section forward of the cockpit. (In its raised position, the nose gave a clean aerodynamic shape at high speed.) The engines required sophis-

ticated control systems in intake and exhaust passages to "tune" airflow for the differences between low-speed and supersonic flight.

Concorde 002 made her maiden flight from Filton in Bristol to Fairford in Gloucestershire on April 9, 1969, piloted by Brian Trubshaw. More than 400 test flights were made. A speed of Mach 1 was reached on March 25, 1970, with Mach 2 being reached on November 12th of that same year.

By 1971, its developers were sufficiently confident of Concorde 002's performance to undertake a demonstration tour. In June, the aircraft visited the Middle East, the Far East and Australia, flying 45,000 miles and visiting twelve countries in thirty days. It was a remarkable achievement, and Concorde 002 created tremendous interest wherever it went.

However, serious concerns about cost, commercial viability, sonic booms and engine noise were growing. It soon became apparent that most national governments would not allow supersonic flights over their territory. The dream of fleets of supersonic aircraft circling the earth and bringing continents and peoples closer together foundered on these objections, and in the end, a total of only twenty Concordes were built. Of these, six were used for trials, and fourteen entered airline service (seven with Air France, and seven with British Airways).

To some, the development of Concorde was "one of the worst investment decisions in the history of mankind" (David Edgerton). Others viewed Concorde as a pioneering European venture in aircraft construction, one that advanced British industry to a new level of technological competence. Whatever the conclusion of history may eventually be, the development of the aircraft was a work of enormous technical virtuosity. Before the Concorde fleet was retired in October 2003, its airliners regularly crossed the Atlantic in three hours and fifty minutes, far above the weather and at twice the speed of sound.

—Peter Craig

ROLLS-ROYCE RB211 (1970)

When it was developed during World War II, the jet engine was seen as a power plant for fast military aircraft. The exhaust of a jet engine consists of a very high-speed airflow, and if this is used to power a correspondingly high-speed aircraft, that aircraft's efficiency is enhanced. However, when powering a relatively low-speed aircraft (such as a conventional airliner), the propulsive efficiency is low, and energy is wasted. The reason is that exhaust leaves the jet engine with excess speed relative to the surrounding air, and its energy is then dissipated in the atmosphere by turbulent mixing—the source of the tremendous noise associated with an older generation of jet aircraft.

Contrary to expectations, the jet engine was adapted for civil use in the years shortly after the war. Initially, its inefficiency for this use did not seem to matter, because new passenger jets such as the de Havilland Comet and the Boeing 707 offered a great improvement in smoothness, internal noise levels, and speed. However, as fuel costs rose and as the public opposition to noise at airports grew, there was growing pressure on aircraft-engine manufacturers to tackle the old problems.

Fortunately, the solution to the problems was the same, and had been proposed by, among others, the gas-turbine pioneer A.A. Griffith in the 1930's (*see* **Flying Bedstead Test Rig**). The principle was to adapt the gas turbine to produce a high volume of relatively slow-moving air.

In such a high-bypass-ratio design, the core engine takes only a small proportion (perhaps a quarter) of the engine's total air intake and serves as a workhorse to drive the rest of the air around the engine with a multi-blade ducted fan. The apparent drawback of such a scheme would be the air resistance and weight of the large nacelles that would house the fan. The Rolls-Royce performance engineers who oversaw the design and installation of engines in commercial aircraft believed that these problems would cancel out gains in propulsive efficiency.

But, as the originator of the RB211 concept, Geoffrey Wilde (1917–2007) had made perfor-

Rolls-Royce RB211 and Olympus 593 Mk 3B engines in the Science Museum, London. The Olympus engine (foreground) was made for the prototype Concorde 002 in the 1970's. The RB211 was conceived because of the problems associated with noise and rising fuel costs.

mance calculations that indicated to him that a very high bypass ratio would work, and he could not understand the opposition to it, recalling that, "after all, they could do the same sums as me." Nevertheless, Rolls-Royce became committed to very high bypass ratios only when it became clear that American aircraft builders, particularly Boeing, were convinced that it was the correct technical path to follow for the coming generation of widebody "jumbo" airliners. The general architecture of the RB211 began to take shape in 1967.

The development problems faced by Rolls-Royce were immense. The size of many components was more than twice that of anything the company had built before. And, to obtain the required efficiency from the core engine, operating temperatures had to be higher than they had been in previous Rolls-Royce engines. In addition, a decision had been made to make the large fan out of Hyfil, which was itself a new carbon-fiber composite material.

In 1970, it became clear that the program was in trouble. The Hyfil fan blades, although they were

immensely strong in tension, could withstand neither the mandatory bird-strike test nor rainwater erosion, thus necessitating a redesign—this time in titanium. This was one of numerous problems; development was proving more difficult and more costly than anyone had foreseen.

Stanley Hooker (1907–84), the designer of the Merlin supercharger (*see* **The Rolls-Royce Merlin**) during World War II, was brought in to take charge of the engineering effort. Rolls-Royce then rededicated itself to the task it had always performed best: meticulous development. However, the project was hugely over budget, and though the unproven engine was almost ready, Rolls-Royce became insolvent in 1971. Fortunately for the RB211, the company was rapidly nationalized by the British government, which agreed to finance the day-to-day costs of the project. Deliveries of the engine began in 1972, and it has since become a great success.

The launch of the RB211 was a unique event in aircraft-engine history. There had been a rapid evolution in jet-engine development in the years after World War II, with substantial leaps in performance coming with each new generation of engine. However, big-fan engines such as the RB211 may prove to be the last to be developed from scratch—as leaps into the unknown. Today, aircraft-engine design is mature, and performance gains are derived from incremental improvements to a basic design—a less risky operational strategy.

The RB211 has proven to be a crucial engine for Rolls-Royce. It has provided the basis for subsequent development of all of the company's large engines. Without a big-fan engine suitable for aircraft such as the Boeing 747, the company would not have remained a major engine manufacturer. Ironically, the engine that bankrupted the most famous British engineering company also placed that company in its current position at the forefront of aircraft-engine manufacturing.

—Andrew Nahum

THE FIRST BRAIN SCANNER (1971)

The brain scanner from Atkinson Morley's Hospital in Wimbledon, London, England was the first to be made; it was the experimental model with which the earliest trials on patients were undertaken. Techniques developed with this machine established computed tomography (CT) scanning as a key imaging technology, particularly as a tool for studying the brain. CT scanning is a method that produces an image of a "slice" of the body, using a computer program to reconstruct a picture from data derived from a series of precision x-ray exposures.

From the formal announcement of the device in 1972, the CT scanner was a runaway success, and by 1977, there were 1,130 machines in countries around the world. A large part of this success can be traced to the fact that it provided diagnostic information that was not available by any other established technique. A conventional x-ray picture is an image produced directly on a photographic film by x-rays that are absorbed to different extents according to the different densities of the tissues that they pass through. That means that in an x-ray picture of parts of the body that contain many organs, images of all of them are laid on top of each other, making the images difficult for radiologists to interpret. From the earliest days of radiology, scientists devised techniques to overcome this problem and to reveal one organ against another— for example, by asking patients to swallow a barium meal to reveal the contours of the stomach separately from other organs of the abdomen.

Tomography, originally developed in the 1920's, used a mechanical method of taking pictures of particular layers of the inside of the body. It worked by moving the x-ray tube and plate in opposite directions while the x-ray was being taken, causing all the image layers but the one in the center to become blurred. Most tomographic x-ray machines produced pictures of layers parallel to the body's length. A method for producing x-rays of "slices" across the body was developed in the 1940's, but it was never really anything more than a scientific curiosity.

It was due to developments in computing—not in radiology—that led to a clinically useful form of tomography. The inventor of the first brain scanner, Godfrey (later Sir Godfrey) Hounsfield (1919–2004), had previously worked on computer-related projects at the British company EMI, including the EMIDEC 1100, the first British transistorized business computer. The CT scanner came out of a project he had undertaken to investigate the practical applications of the mathematical theory of image reconstruction, which dates back to 1917 and also has applications in crystallography, microscopy, and radio astronomy.

Initially, Hounsfield conceived of the scanner as a tool of mass screening, to detect cancer in the same way that mass miniature x-rays had been used since the 1940's for detecting tuberculosis. Officials at the British Department of Health suggested that the brain might be a better subject; it had the advantage that (unlike the abdomen, for example) it could be immobilized for the several minutes that the early scans took. In addition, the brain is difficult to photograph with conventional x-rays because it is encased in the bone of the skull, and higher exposure to x-rays is necessary (more than with other parts of the body); also, except by introducing air or some other contrast medium into the brain, it is difficult to differentiate its different structures.

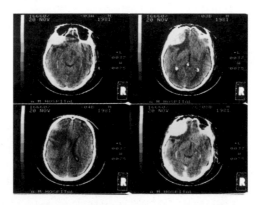

Brain scans taken using the CT scanner
(Atkinson Morley's Hospital)

Hounsfield developed the hardware and software to convert the 160 readings taken from each of the 180 different angles around the head (a total of 28,800) that could be reconstructed on a monitor. One key innovation was to store the readings in a memory device from which the computer would later reconstruct the image in a process that originally took several man-hours. The other essential development was to write a computer program that could produce a medically useful image using the smallest number of calculations. Working from a specialist's knowledge of computer technology on a complex area of mathematical theory, Hounsfield was able to produce an innovative piece of medical technology from outside of its established community.

Although EMI was best known for its entertainment interests, its management had, since World War II, pursued a policy of diversification into both civilian and military electronics projects. Its first venture into medical equipment, the thermographic imager, had not been a notable commercial triumph, but the medical equipment market was still one in which EMI was anxious to become involved. When it became clear that the CT scanner worked and that clinicians were enthusiastic about its diagnostic potential, EMI management made a firm decision to enter the medical market, and even acquired two other electromedical companies to consolidate its position. It has been suggested that without the profits it made from the extraordinary growth of popular music in the 1960's, EMI would never have been able to contemplate going into mass production of the scanner. Yet, as established medical equipment companies (particularly in Japan) began to supply CT scanners in substantial numbers, EMI began to find the competition difficult. This problem coincided with a downturn in its fortunes in the popular music market, and EMI was forced out of the medical equipment business, selling out to General Electric in 1980—only eight years after the scanner was first introduced.

—Tim Boon

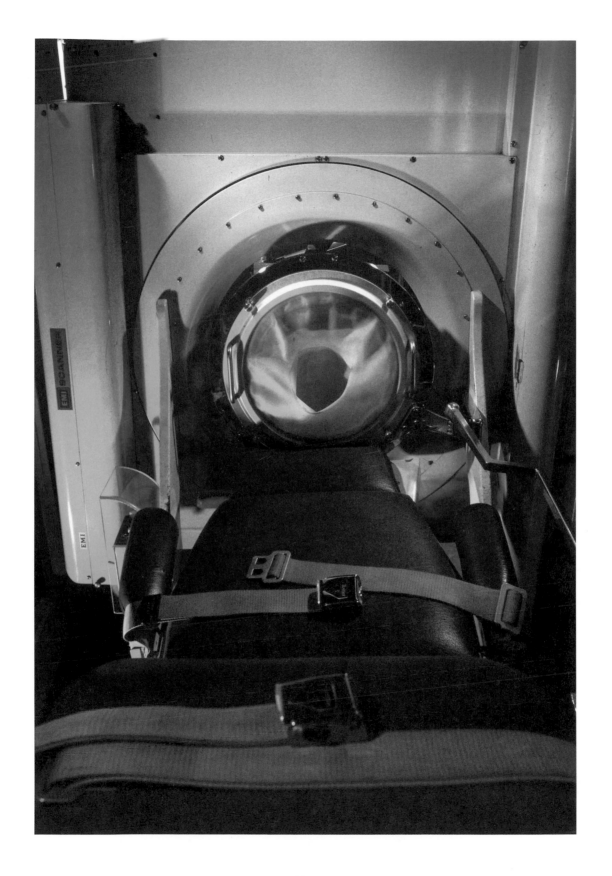

MAGNETIC RESONANCE IMAGING (LATE 1970'S)

Chemists employ nuclear magnetic resonance (NMR) (*see* **Nuclear Magnetic Resonance**) to study hydrogen atoms in many different chemical compounds; it is the variation of the proton signal in these different chemical surroundings that is of interest. NMR can also be used to study the signal of the hydrogen atoms of one ubiquitous compound: water. When placed in a magnetic field, water emits an NMR signal; in fact, this signal can be used to measure the strength of the earth's magnetic field. The Ph.D. student who developed this magnetometer, Peter (later Sir Peter) Mansfield (born 1933), later studied the NMR signals from a much more complex water container: the human body.

The origins of magnetic resonance imaging (MRI)—the use of NMR in medicine—arose from the desire of medical researchers to detect cancerous tumors at an early stage, when they could be treated more easily. In the late 1960's, a young American medical professor, Raymond Damadian (born 1936), who had accepted the theory that cancer cells swell up with water, wondered if NMR could show this phenomenon. By putting cancer cells in an NMR machine, he was able to show that the way cancer cells stored water was indeed different from the way in which normal cells did—but his hope that NMR could be used to characterize cancer cells at an early stage was later shown to be unfounded, as the underlying theory was incorrect. Meanwhile, an American NMR chemist, Paul C. Lauterbur (1929–2007), was trying to improve the sensitivity of NMR in chemistry, but wasn't having much success. But, he learned about Damadian's work and realized that if he placed a magnetic gradient across the otherwise uniform magnetic field, it would be possible to differentiate between various parts of the body that were being irradiated with radio waves. At a stroke, he had converted Damadian's diagnostic method for observing cancer into an imaging technique for the entire human body.

As is often the case, however, there was an enormous gap between this conceptualization and a viable medical technique. Three British groups played an important role in the development of MRI. The first group to enter the field, based at the University of Nottingham, was headed by Peter Mansfield. Mansfield and his group solved the problem of obtaining an image of a cross-section of a human body—one of their first images was a cross-section of a scallion. There was also another group at the University of Nottingham; it was led by E. Raymond Andrew (1921–2001), a pioneer of solid-state NMR who had worked in the United States in the late 1940's. Andrew looked at the ways in which radio frequencies penetrated the human body. He carried out these experiments on small animals, but in 1978, he captured an image of a human forearm.

Andrew and Mansfield were both physicists, but the third British group, at the University of Aberdeen in Scotland, was led by a physician, John Mallard. Mallard had originally been interested in using, for medical purposes, another chemical-physical technique, electron spin resonance—but he switched to NMR after reading Damadian's early papers, and he was already on the way to developing an imaging technique when Lauterbur's findings were published. As a physician, Mallard quickly built up clinical experience with the technique, but he fell behind technically. He also ran into funding problems with the U.K.'s Medical Research Council, whose members felt that a grant for one MRI machine was sufficient.

MRI images are a product of computer-based mathematical analysis; unlike x-ray images, they do not produce direct images of the body. The computer-generated image is a result of the specific MRI procedure selected by the operator. By the late 1970's, research groups had produced remarkably clear images of the inside of the body, and businesses started to move into the field with the aim of bringing the technique to the marketplace. An important factor in this interest was the previously commercially successful development of CT scanning. The companies involved were either the successful developers of commercial CT scanners or those that had come late into the field and were seeking an alternative. These companies came from all over the world; although British pioneers played a significant role in the development of MRI, its commercialization has not been led by British firms.

The benefits of MRI have been both momentous and extensive. Its contributions to the field of medical diagnostics and research have been many: the inner structures of the human body are now able to be non-invasively explored—without the use of ionizing radiation; pain, discomfort and risk for patients have been minimized; the 3-D images offered by an MRI are now vital in much pre-surgical planning; and afflictions in all parts of the body—especially cancers and brain and spinal cord maladies—are now far easier to detect.

For their "discoveries concerning magnetic resonance imaging," both Paul Lauterbur and Sir Peter Mansfield were awarded the Nobel Prize in Physiology or Medicine in 2003.

—Peter Morris

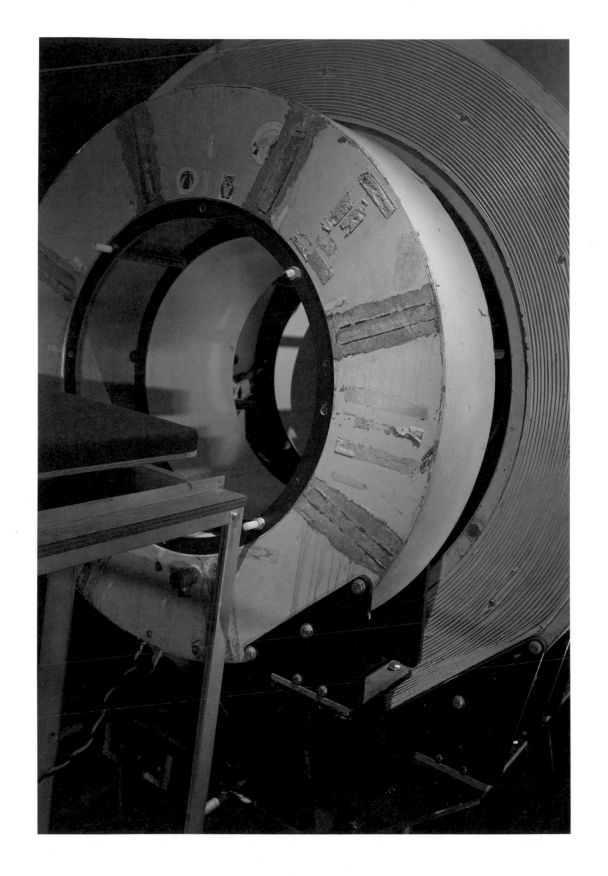

DAVID HOCKNEY:
NATIONAL MEDIA MUSEUM PHOTO COLLAGE (1985)

David Hockney (born 1937) has had a lifelong love-hate relationship with photography. His father was a devoted amateur photographer, and young Hockney first used a box camera on family vacations. As a student and young artist, he used photographs as reference material for his paintings—both those photographs he found in magazines and, increasingly, those he took himself. For Hockney, such photographs were a means to his artistic ends rather than art themselves; the camera was a convenient recording device. Painting and drawing, where the work of the artist's hand can be clearly traced, were what he really loved.

That attitude began to change in 1969 when he and a friend, Peter Schlesinger, photographed each other sitting at opposite ends of a Paris park bench. Later, Hockney stuck the two snapshots together so that the two men appeared together in the final picture. He had made his first "joiner" (as he was to call his later, more complicated collages), and, with very simple means, achieved three complex results.

First, he had made two people sit together when they had not actually done so. Second, by photographing from two slightly different spots, he had shifted his viewpoint and distorted the perspective, yet in a way that is not at all disturbing to the viewer. Finally, and most tellingly, he had suggested something in purely visual terms about the vulnerable, risky nature of a relationship between people who are at once together and apart.

Hockney perceived that "some of the photographs intended to become paintings didn't necessarily have to become paintings," and he embarked on a series of joiners; in this case, rectangular collages of small Polaroid prints. Containing dozens, even hundreds, of Polaroids—painstakingly shot from different viewpoints, closing in on some details, stepping back for others—they were carefully assembled into rectangular grids of white-bordered, square prints. The results were complicated, fascinating and fragmented.

When Hockney progressed to conventional snapshot cameras, the Polaroids looked almost like five-finger exercises—preparations for the real thing. Now, he was able to work more quickly and to break out of the conventional rectangular picture frame, which he had always found inhibiting. The joiners became increasingly complex in construction, scope, ambition, and, most importantly, in their use of time. Prowling restlessly about the people or places that were his subjects, the artist looked at them over an extended period of time. Nothing could be further from our idea of a photograph—which we are naturally inclined to believe is the work of a split-second. But, in a Hockney joiner, nothing and nobody is quite the same when the picture is finished as they were when it had begun.

In June 1985, Hockney accepted an invitation from the National Media Museum in Bradford, Yorkshire to make a joiner of some subject in or around the Museum. He chose the building itself, seen from the same viewpoint as the rather romantic picture of it used by the City of Bradford in its publicity.

The whole exercise took three days, with the actual photography being done, inevitably, in public (several members of the Museum staff and the press appear in the final picture). Hockney and his assistant moved into the Museum to assemble the prints as they were returned by C.H. Wood Ltd., the long-established Bradford photographers and processors. At each session, twenty-five visitors were invited to watch the artist at work and to share his toil as he created.

The picture grew more and more complex, and Hockney was still taking photographs three days after he began. By then, Wood had closed for the weekend, and the final negatives had to be sent to another laboratory. The fact that the colors in these prints were far bluer than in the previous ones did not put off Hockney. Placed on the fringes of the picture, they gave it some of the colors of his California home and added to the richness of the final result.

Hockney has always been fascinated by new technologies, and, when fax machines and color laser printers came on the market, he added them to his arsenal, using them in novel and personal ways. As well, in 1990, he acquired a still video camera, which he brought to the Museum in 1991, exhibiting his electronic still portraits and giving lectures about them.

The electronic stills took even further Hockney's attempts to overcome what he saw as two major drawbacks of photography. First, the one-eyed view of the world usually imposed by a camera lens—which Hockney seeks to avoid as his multi-perspective journey down each person's body begins to approach what one might see with one's own eyes; second, the "air" between lens and subject—which Hockney tries to eliminate by bringing the camera so close to his subject. After seeing Hockney's "direct confrontations," ordinary color photographs could look plastic, dull-toned, and too blue.

Hockney's use of technology in his art continues to this day; he has, in 2009, unveiled iPhone artworks, and, in 2010, started to use the iPad and its stylus as a sort of digital sketch pad for many of his musings. Hockney's work has always been colorful, inventive and witty. His own words describe his aims better than do anyone else's descriptions: "I do want to make a picture that has meaning for a lot of people. I think the idea of making pictures for twenty-five people in the art world is crazy and ridiculous. It should be stopped…"

—COLIN FORD

GENETICALLY ENGINEERED MICE (1988)

London's Science Museum collects artifacts, not organisms. This rule has applied at the Museum since its founding. But in 1989, the rule was broken when two mice were acquired for its permanent collections. The mice were male, their bodies had been preserved by freeze-drying, and they were the gift of the Harvard Medical School.

The interest in these mice reflects the revolution in the biological sciences that has accompanied the development of what is often termed "genetic engineering." Following Francis Crick and James Watson's discovery of the structure of the genetic material DNA in 1953 (*see* **Crick and Watson's DNA Model**), molecular geneticists made astonishingly rapid progress in unraveling the mechanism of inheritance. By the late 1960's, they knew a great deal about how genetic information is passed down through generations, and they had worked out how this information is translated into specific proteins. In the early 1970's, they acquired the ability to perform a sort of molecular surgery with DNA. Known as "recombinant DNA technology," it involved extracting DNA fragments from one organism and inserting them into the genetic structure of another. Over the past forty years, recombinant DNA technology has transformed molecular genetics from an elegant but essentially pure science into a powerful applied science that sits at the heart of biotechnology and molecular

medicine. Applications include the creation of new crop varieties, the development of new drugs, and the production of hormones, such as human insulin, on an industrial scale.

One particularly important branch of recombinant DNA technology involves the creation of so called transgenic animals. A transgenic animal is produced when donor DNA fragments are inserted into a newly fertilized egg cell and become incorporated into the animal's genes. As it develops, the inserted DNA is copied along with the animal's own genes, and the result is a transgenic animal with new characteristics. In 1984, Timothy Stewart, Paul Pattengale and Philip Leder of the Harvard Medical School's Department of Genetics announced the creation of thirteen strains of transgenic mice. These mice had been modified by the insertion of a part of a cancer-causing gene—an oncogene. As a result, the mice were predisposed to develop specific cancers. By working with an "OncoMouse," medical researchers could explore the underlying mechanism of cancer and test potential anti-cancer drugs.

Harvard University applied for a patent on Stewart and his colleagues' OncoMouse in 1984. A precedent for this patent application had been set in 1980, with the U.S. Supreme Court ruling that "anything under the sun that is made by man" is patentable. In 1987, the U.S. Patent Office

announced that "higher life forms" were patentable, and in 1988 it finally granted Harvard's OncoMouse patent application. In 1991, the European Patent Office followed its American counterpart by granting a European patent on the Harvard Onco-Mouse. Eventually, the OncoMouse was marketed by the DuPont chemical company, and initially sold for about $100—as compared with around one dollar for an ordinary laboratory mouse. The Onco-Mouse—not without controversy—has been widely used in cancer research, and other mouse varieties have been developed as models for studying heart disease and other genetically inherited diseases.

The patenting of life forms is a divisive issue. Some animal welfare and environmental organizations oppose plant and animal patents on ethical grounds; some farming organizations oppose them on commercial grounds. Such objections notwithstanding, the rapid development and commercial application of genetic biotechnologies have provided great scope for plant and animal patents. The Harvard OncoMouse thus represents an important phase in the development of molecular genetics. With the advent of biotechnology and transgenic animals, it seems that organisms can also be artifacts.

—John Durant

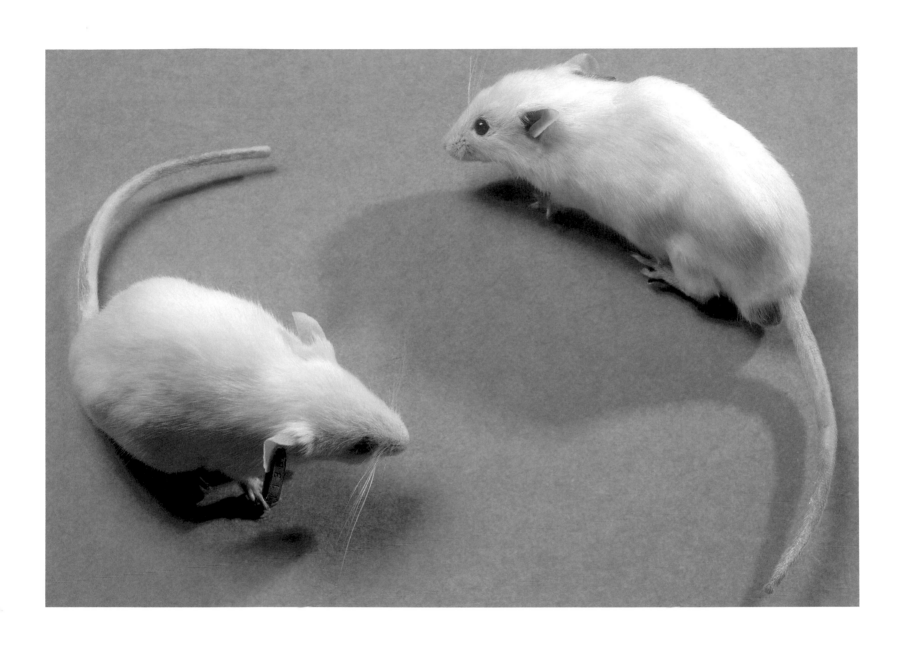

PROTOTYPE "CLOCK OF THE LONG NOW" (1999)

"Civilization is revving itself into a pathologically short attention span. The trend might be coming from the acceleration of technology, the short-horizon perspective of market-driven economics, the next-election perspective of democracies, or the distractions of personal multitasking. All are on the increase." This analysis of society at the end of the 20th century was written in 1998 by Stewart Brand (born 1938), writer, inventor and founder of the *Whole Earth Catalog*. Brand, together with computer designer Danny Hillis (born 1956) and other prominent *fin de siècle* thinkers, had become increasingly concerned that the year 2000 had come to be seen as a temporal mental barrier to the future. Brand explained: "Some sort of balancing corrective to the short-sightedness is needed—some mechanism or myth that encourages the long view and the taking of long-term responsibility, where 'the long term' is measured at least in centuries."

Hillis' proposal was to build "both a mechanism and a myth," a monumental-scale mechanical clock capable of telling time for 10,000 years—if it was maintained properly. Such a clock would prompt conversations about "deep time," perhaps becoming a public icon for time in the same way that photographs of earth from space taken by the Apollo 8 crew in December 1968 have become icons for a fragile planet in boundless space. (It was partly due to Brand's agitation that NASA released earlier satellite-based photographs of earth to the public in 1966.)

In 1996, Brand and Hillis formed a board of like-minded friends. Calling themselves "The Long Now Foundation," the organization's title sprang from a suggestion by musician and composer Brian Eno that "The Long Now" could be seen as an important extension of human temporal horizons. In this scheme, "now" was seen as the present moment plus or minus a day, and "nowadays" extended the time horizon to a decade or so forward and backward. However, the "long now" would dramatically extend this "time envelope." Since

settled farming began in about 8000 B.C., the futurist Peter Schwartz proposed that the "long now" should mean the present day plus or minus 10,000 years—"about as long as the history of human technology," explained Hillis.

The design principles established for the clock laid down strict parameters for its construction. With occasional maintenance, it was thought that the clock should reasonably be expected to display the correct time for 10,000 years. It was designed to be maintainable with Bronze Age technology. The plan was also that it should be possible to determine the operational principles of the clock by close inspection, to improve the clock over time and to build working models of the clock from table-top to monumental size using the same design.

In 1997, a small team of expert engineers, mechanics and designers based in San Francisco, led by Alexander Rose, set about constructing a prototype of the Clock of the Long Now, as the project became known. Driven by the power of two falling weights, which are wound every few days, the torsional (twisting) pendulum beats twice per minute, transmitting its time through an oversized watch-escapement mechanism to the heart of the clock, a mechanical computer. This computer, conceptually linked to the machines of 19th-century polymath Charles Babbage (*see* **Babbage's Calculating Engines**), operates once every hour, updating timekeeping elements within the dial display, including the position of the sun, the lunar phase and the locally-visible star field. The slowest-moving part of this display indicates the precession of the equinoxes. As the designer of some of the world's fastest supercomputers in the 1980's, Danny Hillis said in the 1990's that he wished to "atone for his sins" of speeding up the world by designing the world's slowest computer for the Clock of the Long Now.

This range of tempos reflects the Foundation's idea of "layers of time" in human existence. The fastest-changing layer is fashion and art; a little

slower is commerce. Infrastructure and governance take still longer to change. Cultures change very slowly, with nature reflecting the slowest tempo of all. "The fast layers innovate; the slow layers stabilize," explained Brand. The Foundation believes that an understanding of the opportunities and threats embodied in these layers of temporal change is crucial in correcting mankind's apparent short-sightedness. These ambitions and ideals were expressed eloquently in the finished prototype clock, which first ticked in San Francisco moments before the end of New Year's Eve 1999. It was then moved to London, where it had been selected as the final exhibit in the Science Museum's Making the Modern World gallery, opened by Queen Elizabeth II in 2000.

Meanwhile, the Foundation continued to build further prototypes, refining the design of the clock's several constituent subassemblies in preparation for the construction (now underway) of a 10,000-year clock inside a mountain in western Texas, near the town of Van Horn. The Foundation hopes to build several "millennial clocks" over the course of time, and a site for another has been purchased atop a mountain in eastern Nevada, adjacent to Great Basin National Park.

By its nature, the clock is both a conclusion—of a long process of human thinking, making and acting (as represented in the range of inventions covered by the essays in this book)—and a starting point, for a long future, the contents of which are uncertain, the opportunities of which are infinite. Stewart Brand observed, "This present moment used to be the unimaginable future." As a symbol for the past, present and future of human ingenuity, the Clock of the Long Now is a fitting device to represent the modern world and all of its milestones. As Danny Hillis has said, "Time is a ride—and you are on it."

—DAVID ROONEY

CONTRIBUTORS

Most contributors are or have been senior staff members of Great Britain's National Museum of Science and Industry (NMSI), comprising the Science Museum, London; the Science Museum, Swindon; the National Railway Museum, York; the Fleet Air Arm Museum, Yeovilton; and the National Media Museum, Bradford. The NMSI positions they either held or currently hold are given with their names:

Marcus Austin
Research Assistant, Special Products Group

John Bagley
Senior Curator, Aeronautics

Peter Bailes
Research Assistant, Special Products Group

John Becklake
Head of the Technology Group

Tilly Blyth
Curator, Computing and Information, Science Museum

Tim Boon
Curator, Public Health

Brian Bowers
Senior Curator, Electrical Engineering

Roger Bridgman
Curator, Communications

Neil Brown
Senior Curator, Classical Physics

Robert Bud
Head, Life and Environmental Sciences Group

Jane Bywaters
Manager, Interpretation Unit

Sue Cackett
Curator, Materials Science

Janet Carding
Research Assistant, Special Projects Group

Vicky Carroll
Curator, History of Science, Science Museum

John Coiley
Head, National Railway Museum, York

Neil Cossons
Director, National Museum of Science and Industry

Peter Craig
Curator, Concorde Exhibition, Fleet Air Arm Museum, Yeovilton

John Darius
Senior Curator, Astronomy & Mathematics

Eryl Davies
Research Assistant, Special Projects Group

John Durant
Assistant Director, Science Communication Division

Stuart Emmens
Collections Assistant, Public Health

Graham Farmelo
Head of Education and Interpretation

Colin Ford
Head, National Media Museum, Bradford

Graeme Fyffe
Collections Development Librarian, Science Museum Library

John Griffiths
Head, Special Projects Group

Colin Harding
Kodak Curator, National Media Museum, Bradford

Alex Hayward
Research Assistant, Special Projects Group

Christine Heap
Curator, Information & Support Services, National Railway Museum, York

Dieter Hopkin
Curator, Collections, National Railway Museum, York

Jane Insley
Curator, Environmental Sciences

Kevin Johnson
Collections Assistant, Astronomy & Mathematics

Stephen Johnston
Research Assistant, Special Projects Group

Jane Kirk
Registrar

Ghislaine Lawrence
Senior Curator, Clinical Medicine

John Liffen
Collections Assistant, Road & Rail Transport

Peter Mann
Senior Curator, Road & Rail Transport

Doug Millard
Senior Curator, ICT and Space Technology, Science Museum

Peter Morris
Principal Curator of Science

Alan Morton
Senior Curator, Modern Physics

Sue Mossman
Research Assistant, Special Projects Group

Graham Mottram
Deputy Director and Curator, Fleet Air Arm Museum, Yeovilton

Andrew Nahum
Senior Curator, Astronautics

Ann Newmark
Senior Curator, Documentation

Francesca Riccini
Research Assistant, Special Projects Group

Pippa Richardson
Interpretation Officer

Derek Robinson
Head, Science Group

John Robinson
Senior Curator, Water Transport

David Rooney
Curator, Transport

Tony Sale
Manager, Computer Conservation Programme

Ross Sharp
Aviation Operators Officer, Science Museum, Wroughton

Wendy Sheridan
Curator, Pictorial Collection

Rob Shorland-Ball
Deputy Head, National Railway Museum, York

Peter Stephens
Curator, Civil Engineering

Doron Swade
Senior Curator, Computing

Roger Taylor
Senior Curator, Photography, National Media Museum, Bradford

John Trenouth
Senior Curator, Television, National Media Museum, Bradford

Peter Turvey
Research Assistant, Special Projects Group

Rod Varley
Senior Curator, Film, National Media Museum, Bradford

Denys Vaughan
Senior Curator, Time Measurement

Jane Wess
Curator, Astronomy & Mathematics

Anthony Wilson
Publications Manager

David Woodcock
Collections Assistant, Electrical Engineering

Michael Wright
Curator, Mechanical Engineering

Tom Wright
Assistant Director, Collections Division

The primary images of Science Museum artifacts accompanying the essays were specially commissioned from photographer **Philip Sayer**.

FURTHER READING

BYZANTINE SUNDIAL-CALENDAR

Field, J.V., "Some Roman and Byzantine Portable Sundials and the London Sundial-Calendar," *History of Technology*, Vol. 12 (1990): pp. 103–35

Field, J.V. and M.T. Wright, "Gears from the Byzantines," *Annals of Science*, Vol. 42 (1985): pp. 87–138

Hill, D.R., "Al-Biruni's Mechanical Calendar," *Annals of Science*, Vol. 42 (1985): pp. 139–63

Price, D.J. de S., "Gears from the Greeks," *Transactions of the American Philosophical Society*, new series, Vol. 64, Part 7 (1974)

Wright, M.T., "Rational and Irrational Reconstruction: The London Sundial-Calendar and the Early History of Geared Mechanisms," *History of Technology*, Vol. 12 (1990): pp. 65–102

ISLAMIC GLASS ALEMBIC

Anderson, R.G.W., "Early Islamic Chemical Glass," *Chemistry in Britain*, 1983, p. 822

THE GIUSTINIANI MEDICINE CHEST

Burnett, John, "The Giustiniani Medicine Chest," *Medical History*, Vol. 26, No. 3 (1982): pp. 325–33

Griffenhagen, George B. and Mary Bogard, *History of Drug Containers and Their Labels*, Madison, WI: American Institute of Pharmacy, 1999

THE STANDARDS OF THE REALM

Connor, R.D., *The Weights and Measures of England*, London: HMSO, 1987

Skinner, F.G., *Weights and Measures: Their Ancient Origins and Their Development in Great Britain up to A.D. 1855*, London: HMSO, 1967

Zupko, Ronald Edward, *British Weights and Measures: A History from Antiquity to the Seventeenth Century*, Madison, WI: University of Wisconsin Press, 1977

SLIDE RULE BY ROBERT BISSAKER

Rudowski, Werner H., "The Bissaker Slide Rule in the Science Museum London," *Bulletin of the Scientific Instrument Society*, No. 97 (2008): pp. 34–48

Turner, Anthony, *Early Scientific Instruments: Europe 1400–1800*, London: Sotheby's, 1987, pp. 161–65

NAPIER'S BONES

Bryden, D.J., *Napier's Bones: A History and Instruction Manual*, London: Harriet Wynter, 1992

Cole, Frank, *Proto-Logs: The Complete Logarithms of John Napier (1550–1617): A Realisation*, London: Frank Cole, 1999

Horsburgh, E.M., (ed.), *Handbook of the Exhibition of Napier Relics and of Books, Instruments and Devices for Facilitating Calculation*, Edinburgh: Royal Society of Edinburgh, 1914; reprinted as *Modern Instruments and Methods of Calculation: A Handbook of the Napier Tercentenary Exhibition*, London: G. Bell/Royal Society of Edinburgh, 1916; facsimile edition, *Handbook of the Napier Tercentenary Celebration of Modern Instruments and Methods of Calculation*, Los Angeles: Tomash, 1982

Williams, Michael R., *A History of Computing Technology*, Englewood Cliffs, NJ: Prentice Hall, 1985 (chapter 2)

HAUKSBEE'S AIR PUMP

Hauksbee, Francis, *Physico-Mechanical Experiments on Various Subjects*, London: R. Brugis, 1709; 2nd edition, 1719; reprinted with a new introduction by Duane H.D. Roller, London and New York: Johnson Reprint Corporation, 1970

Heilbron, J.L., *Electricity in the 17th and 18th Centuries: A Study of Early Modern Physics*, Berkeley, CA: University of California Press, 1979

THE ORIGINAL ORRERY

King, Henry C. and John R. Millburn, *Geared to the Stars: The Evolution of Planetarium, Orreries and Astronomical Clocks*, Toronto: University of Toronto Press, 1978

Whiting, Michael, "Orrery Developments: The Use of Meccano in Constructing Planetaria," *Bulletin of the Scientific Society*, Vol. 94 (2007)

SISSON'S RULE

Roy, William, "An Account of the Measurement of a Base on Hounslow-Heath," *Philosophical Transactions of the Royal Society*, Vol. 75 (1785): pp. 385–480

SHELTON'S ASTRONOMICAL REGULATOR

Howse, Derek, "Captain Cook's Pendulum Clocks," *Antiquarian Horology*, Vol. 6 (1969): pp. 62–76

Howse, Derek and Beresford Hutchinson, "The Saga of the Shelton Clocks," *Antiquarian Horology*, Vol. 6 (1969): pp. 281–98

Woolf, Harry, *The Transits of Venus: A Study of Eighteenth-Century Science*, Princeton, NJ: Princeton University Press, 1959

ARKWRIGHT'S SPINNING MACHINE

English, Walter, *The Textile Industry: An Account of the Early Inventions of Spinning, Weaving, and Knitting Machines*, Harlow, Essex: Longman, 1969

Mohanty, Gail Fowler, *Labor and Laborers of the Loom: Mechanization and Handloom Weavers, 1780–1840*, New York: Routledge, 2006

Tann, Jennifer, *The Development of the Factory*, London: Cornmarket Press, 1970

TROUGHTON'S DIVIDING ENGINE

Chapman, Allan, *Dividing the Circle: The Development of Critical Angular Measurement in Astronomy 1500–1850*, Hemel Hempstead, Hertfordshire: Ellis Horwood, 1990; 2nd edition, Chichester, West Sussex and New York: John Wiley, 1995

Ramsden, Jesse, *Description of an Engine for Dividing Mathematical Instruments*, London, 1777

Rolt, L.T.C., *Tools for the Job: A Short History of Machine Tools*, London, Batsford; reprinted as *Tools for the Job: A History of Machine Tools to 1950*, London: HMSO, 1986

Troughton, Edward, "Graduation [in the Making of Mathematical Instruments],"in *The Edinburgh Encyclopedia*, edited by David Brewster, Vol. 10, Edinburgh: Blackwood, 1830, pp. 348–84; Philadelphia: J. and E. Parker, 1832

HERSCHEL'S SEVEN-FOOT TELESCOPE

Herschel, Caroline, *Memoir and Correspondence of Caroline Herschel, by Mrs John Herschel*, London: John Murray, 1876; edited by Mary Cornwallis Herschel, Cambridge and New York, Cambridge University Press, 2011

Herschel, John, *Telescope*, Edinburgh, A. & C. Black, 1860

King, Henry C., *The History of the Telescope*, London: Charles Griffin, 1955; reprinted London and New York: Dover, 1979

Murdin, Paul, *Secrets of the Universe: How We Discovered the Cosmos*, Chicago: University of Chicago Press, 2009

Ogilvie, Mary Bailey, *Women in Science: Antiquity through the Nineteenth Century: A Biographical Dictionary with Annotated Bibliography*, Cambridge, MA: MIT Press, 1986

BOULTON AND WATT ROTATIVE ENGINE

Dickinson, H.W. and Rhys Jenkins, *James Watt and the Steam Engine: The Memorial Volume Prepared for the Committee of the Watt Centenary Commemoration at Birmingham*, Oxford: Clarendon Press, 1927

Farey, John, *The Steam Engine: Historical, Practical and Descriptive*, London: Longman, Rees, Orme, Brown and Green, 1827; reprinted Newton Abbot, Devon: David & Charles, 1971

Hulse, David K., *The Early Development of the Steam Engine*, Leamington Spa, Warwickshire: TEE Publishing, 1999

Law, R.J., *The Steam Engine: A Brief History of the Reciprocating Engine*, London: HMSO, 1965; reprinted 1986

Law, R.J., *James Watt and the Separate Condenser: An Account of the Invention*, London: HMSO, 1976

Tann, Jennifer, "Boulton and Watt's Organisation of Steam Engine Production before the Opening of the Soho Foundry," *Transactions of the Newcomen Society*, Vol. 49 (1978)

SYMINGTON'S MARINE ENGINE

Harvie, Christopher T., John Compton and W.J. Harvey, *The First Practical Steamboat: William Symington and the Charlotte Dundas*, Grangemouth: Scottish Industrial Heritage Society, 2003

Harvey, W.S. and G. Downs-Rose, *William Symington: Inventor and Engine Builder*, London: Northgate, 1980

RAMSDEN'S THREE-FOOT THEODOLITE

Ordnance Survey, *Account of the Observations and Calculations of the Principal Triangulation*, London: Eyre and Spottiswoode, 1858

Seymour, W.A. (ed.), *A History of Ordnance Survey*, Folkstone, Kent: Dawson and Sons, 1980

MAUDSLAY'S SCREW-CUTTING LATHE

Cantrell, John, *Henry Maudslay and the Pioneers of the Machine Age*, Stroud, Gloucestershire: Tempus, 2002

Gilbert, K. R., *Henry Maudslay: Machine Builder*, London: HMSO, 1971

Holtzapffel, Charles, *Turning and Mechanical Manipulation: Intended as a Work of General Reference and Practical Instruction on the Lathe, and the Various Mechanical Pursuits Followed by*

Amateurs, Vol. 2, London: Holtzapffel, 1846; reprinted Lakeville, MN: Astragal Press, 1990

Smiles, Samuel, *Industrial Biography: Iron Workers and Tool Makers*, London: John Murray, 1863; reprinted Newton Abbot, Devon: David & Charles, 1967

HERSCHEL'S PRISM AND MIRROR

Cornell, E.S., "The Radiant Heat Spectrum from Herschel to Melloni: Herschel and His Contemporaries" (Part I), *Annals of Science*, Vol. 3 (1938): pp. 119–37

Herschel, William, "Investigation of the Powers of the Prismatic Colours to Heat and Illuminate Objects," *Philosophical Transactions of the Royal Society*, Vol. 90 (1800): pp. 255–83

Herschel, William, "Experiments on the Refrangibility of the Invisible Rays of the Sun," *Philosophical Transactions of the Royal Society*, Vol. 90 (1800): pp. 284–93

Herschel, William, "Experiments on the Solar and on the Terrestrial Rays that Occasion Heat, Part 1," *Philosophical Transactions of the Royal Society*, Vol. 90 (1800): pp. 293–326

Herschel, William, "Experiments on the Solar and on the Terrestrial Rays that Occasion Heat, Part 2," *Philosophical Transactions of the Royal Society*, Vol. 90 (1800): pp. 437–538

COALBROOKDALE BY NIGHT

Briggs, Asa, *Iron Bridge to Crystal Palace: Impact and Images of the Industrial Revolution*, London: Thames & Hudson in association with Ironbridge Gorge Museum Trust, 1979

Hyde, Ralph, *Panoramania!: The Art and Entertainment of the All Embracing View*, London: Trefoil for the Barbican Art Gallery, 1988

Klingender, Francis D., *Art and the Industrial Revolution*, London: Carrington, 1947; revised and edited by Arthur Elton, London: Evelyn, Adams & Mackay and New York: A.M. Kelley, 1968

McKiernan, Mike, "Philip Jacques de Loutherbourg, *Coalbrookdale by Night* (1801)," *Occupational Medicine*, Vol. 58, Issue 5 (2008): pp. 316–17

Smith, Stuart B., *A View from the Iron Bridge*, Telford, Shropshire: Ironbridge Gorge Museum Trust, 1979

PORTSMOUTH BLOCK-MAKING MACHINERY

Beamish, Richard, *Memoir of the Life of Sir Marc Isambard Brunel*, London: Longman, 1862

Coad, Jonathan, *The Portsmouth Block Mills: Bentham, Brunel and the Start of the Royal Navy's Industrial Revolution*, Swindon: English Heritage, 2005

Cooper, C.C., "The Production Line at Portsmouth Block Mill," *Industrial Archaeology Review*, Vol. 6, No. 1 (1982): pp. 28–44

Gilbert, K.R., *The Portsmouth Block-Making Machinery: A Pioneering Enterprise in Mass Production*, London: HMSO, 1965

Rees, Abraham (ed.), "Machinery for Manufacturing Ships' Blocks," London: Longman, Hurst, Rees, Orme and Brown, *New Cyclopaedia*, 1819

TREVITHICK'S HIGH-PRESSURE ENGINE

Burton, Anthony, *Richard Trevithick: Giant of Steam*, London: Aurum Press, 2000

Dickinson, H.W., *A Short History of the Steam Engine*, Cambridge: Cambridge University Press, 1938

Farey, John, *A Treatise on the Steam Engine: Historical, Practical and Descriptive*, London: Longman, Rees, Orme, Brown and Green, 1827; reprinted Newton Abbot, Devon: David & Charles, 1971

Hodge, James, *Richard Trevithick*, Princes Risborough, Buckinghamshire: Shire, 2003

Ince, L., "Richard Trevithick's Patent Steam Engine," *Stationary Power*, Vol. 1, 1984

Rees, Abraham (ed.), "Steam-Engine," London: Longman, Hurst, Rees, Orme and Brown, *New Cyclopaedia*, 1819

DALTON'S WOODEN ATOMS

Dalton, John, *Foundations of the Atomic Theory*, Edinburgh: William F. Clay, 1893

Greenaway, Frank, *John Dalton and the Atom*, Ithaca, NY: Cornell University Press, 1966

McDonnell, John J., *The Concept of an Atom from Democritus to John Dalton*, Lewiston, NY: Edwin Mellen Press, 1992

Morris, Peter, "Feel the Force," *Nature Physics*, Vol. 2, p. 288 (2006)

Patterson, Elizabeth, *John Dalton and the Atomic Theory: The Biography of a Natural Philosopher*, New York: Doubleday, 1970

Thackray, Arnold, *John Dalton: Critical Assessments of His Life and Science*, Cambridge, MA: Harvard University Press, 1972

THE COMET STEAM ENGINE

Hughson, Martin, *John Robertson, Engineer: Designer of the Comet's Engine*, Glasgow: Bearhead & Neilston Historical Association, 1989

PUFFING BILLY

Marshall, Dendy C.F., *Two Essays in Early Locomotive History*, London: Locomotive Publishing, 1928

Ransom, P.J.G., *The Victorian Railway and How It Evolved*, London: Heinemann, 1990

DAVY'S SAFETY LAMP

Davy, Humphry, "On the Fire-Damp of Coal Mines and on Methods of Lighting the Mines so as to Prevent Its Explosion," Part I, *Philosophical Transactions of the Royal Society*, Vol. 105 (1816)

Davy, Humphry, *On the Safety Lamp for Coal Miners: Some Researches on Flame*, London: R. Hunter, 1818

Hardwick, F.W. and L.T. O'Shea, "Notes on the History of the Safety Lamp," *Transactions of the Institution of Mining Engineers*, Vol. 51 (1915): pp. 548–49

Keys, Thomas E., "Sir Humphry Davy and His Safety Lamp for Coalminers," *Mayo Clinic Proceedings*, Vol. 43 (December 1966)

FARADAY'S CHEMICAL CHEST

Cantor, G.N., David Gooding and Frank A.J.L. James, *Faraday*, London: Macmillan, 1998

James, Frank A.J.L. (ed.), *The Correspondence of Michael Faraday*, London: Institution of Electrical Engineers, 5 volumes, 1991–2008

LISTER'S MICROSCOPE

Bracegirdle, Brian, "Famous Microscopists: Joseph Jackson Lister, 1786–1869," *Proceedings of the Royal Microscopical Society*, Vol. 22 (September 1987): pp. 273–97

Deiman, Johannus Cornelis, *Microscope Optics and J.J. Lister's Influence on the Development of the Achromatic Objective, 1750–1850*, Amsterdam: J.C. Deiman, 1992

Fisher, Richard B., *Joseph Lister 1827–1912*, London: Macdonald and Jane's, and New York: Stein and Day, 1977

BELL'S REAPER

Bell, Patrick, "Some Account of Bell's Reaping Machine," *Quarterly Journal of Agriculture*, (1854): pp. 185–95

Fussell, G.E., *The Farmer's Tools, 1500–1900: The History of British Farm Implements, Tools and Machinery Before the Tractor Came*, London: Melrose, 1952; reprinted London: Bloomsbury, 1985

Jones, L.J., "The Early History of Mechanical Harvesting," *History of Technology*, Vol. 4 (1979): pp. 101–48

STEPHENSON'S *ROCKET*

Bailey, Michael R. and John P. Glithero, *The Engineering and History of Rocket: A Survey Report*, York: National Railway Museum, 2000

Bailey, Michael R. and John P. Glithero, *Stephenson's Rocket: A History of a Pioneering Locomotive*, London: Science Museum, 2002

McGowan, Christopher, *Rail, Steam and Speed: The "Rocket" and the Birth of Steam Locomotion*, New York: Columbia University Press, 2004

Reed, Brian, *Locomotives in Profile*, Vol. 1, London: Profile and New York: Doubleday, 1971

Thomas, R.G.H., *The Liverpool & Manchester Railway*, London: Batsford, 1980

BABBAGE'S CALCULATING ENGINES

Bromley, Allan G., "Difference Engines and Analytical Engines," in *Computing Before Computers*, edited by William Aspray, Ames, IA: Iowa State University Press, 1990

Spufford, Francis and Jenny Uglow, *Cultural Babbage: Technology, Time and Invention*, London and Boston: Faber & Faber, 1996

Swade, Doron, *Charles Babbage and His Calculating Engines*, London: Science Museum, 1991

Swade, Doron, *Charles Babbage's Difference Engine No. 2: Technical Description*, London: National Museum of Science and Industry, 1996

Swade, Doron, *The Cogwheel Brain: Charles Babbage and the Quest to Build the First Computer*, London: Little Brown, 2000; as *The Difference Engine: Charles Babbage and the Quest to Build the First Computer*, New York: Penguin, 2001

TALBOT'S *LATTICED WINDOW*

Arnold, H.J.P., *William Henry Fox Talbot: Pioneer of Photography and Man of Science*, London: Hutchinson, 1977

Buckland, Gail, *Fox Talbot and the Invention of Photography*, Boston: Godine, 1980

Schaaf, Larry J., *Out of the Shadows: Herschel, Talbot and the Invention of Photography*, New Haven, CT: Yale University Press, 1992

Talbot, William Henry Fox, *The Pencil of Nature*, London: Longman, Brown, Green and Longmans, 1844; facsimile edition with a new introduction by Colin Harding, Chicago: KWS Publishers, 2011

COOKE AND WHEATSTONE'S TELEGRAPH

Bowers, Brian, *Sir Charles Wheatstone, FRS, 1802–1875*, London: HMSO, 1975; 2nd edition, London: Institute of Electrical Engineers, 2001

Hubbard, Geoffrey, *Cooke and Wheatstone and the Invention of the Electric Telegraph*, London: Routledge & Kegan Paul, 1965; reprinted, 2008

THE BROUGHAM

Hooper, G.N. "Transition in London Carriages," paper read at the Institute of British Carriage Manufacturers, January 23, 1896

EARLY DAGUERREOTYPES OF ITALY

Gernsheim, Helmut, *The Origins of Photography*, London and New York: Thames & Hudson, 1982

THE ROSSE MIRROR

King, Henry C., *The History of the Telescope*, London: Charles Griffin, 1955; reprinted London: Dover, 1979

The Illustrated London News, September 9, 1843: pp. 165–66

The Illustrated London News, April 19, 1845: pp. 253–54

Rosse, William Parsons, Earl, *Account of a New Reflecting Telescope*, Edinburgh: Blackwood, 1828

Rosse, William Parsons, Earl, "Observations on the Nebulae," *Philosophical Transactions of the Royal Society*, (1850): pp. 499–514

Tubridy, Michael, *Reconstruction of the Rosse Six Foot Telescope*, Ferbane, County Offaly, Ireland: Brown Press, 1998

ELIAS HOWE'S SEWING MACHINE

Cooper, Grace Rogers, "The Invention of the Sewing Machine," Washington, D.C., *United States National Museum Bulletin*, No. 254, Smithsonian Institution, 1968; as *The Sewing Machine: Its Invention and Early Development*, New York: Brazillier, 1976

Levitt, Sarah, *Victorians Unbuttoned: Registered Designs for Clothing, Their Makers and Weavers, 1839–1900*, London: Allen & Unwin, 1986

JOULE'S PADDLE-WHEEL APPARATUS

Cardwell, D.S.L., *From Watt to Clausius: The Rise of Thermodynamics in the Early Industrial Age*, London: Heinemann, and Ithaca, NY: Cornell University Press, 1971

Cardwell, D.S.L., *James Joule: A Biography*, Manchester: Manchester University Press and New York: St Martin's Press, 1989

Joule, James Prescott, *The Scientific Papers of James Prescott Joule*, London: Physical Society, 1884–87; reprinted London: Dawsons, 1963

Steffens, Henry John, *James Prescott Joule and the Concept of Energy*, New York: Science History Publications and Folkestone, Kent: Dawson, 1979

ORIGINAL MAUVE DYE

Farber, Edward (ed.), *Great Chemists*, New York: John Wiley, and London: Interscience, 1961, pp. 758–72

Garfield, Simon, *Mauve: How One Man Invented a Colour that Changed the World*, London: Faber & Faber, 2000; New York: Norton, 2002

Gillispie, Charles Coulston (ed.), *Dictionary of Scientific Biography*, Vol. 10, New York: Scribner, 1974, pp. 515–17

Leaback, D.H., "Perkin's Pioneering Enterprise," *Chemistry in Britain*, 1988, pp. 787–90

Obituary Notices, "William Henry Perkin," *Journal of the Chemistry Society*, Vol. 93 (Part II), 1908, pp. 2214–57

Travis, Anthony S., "Perkin's Mauve: Ancestor of the Organic Chemical Industry," *Technology & Culture*, Vol. 31 (1990): pp. 51–82

Travis, Anthony S., *The Rainbow Makers: The Origins of the Synthetic Dyestuffs Industry in Western Europe*, Bethlehem, PA: Lehigh University Press and London: Associated University Presses, 1993

LEWIS CARROLL'S PHOTOGRAPHS

Ford, Colin, *Lewis Carroll, Photographer*, Bradford, Yorkshire: National Museum of Photography, 1987

Gernsheim, Helmut, *Lewis Carroll, Photographer*, London and New York: Dover, 1969

Norton Cohen, Morton, *Reflection in a Looking Glass: A Centennial Celebration of Lewis Carroll, Photographer*, New York: Aperture, 1998

Taylor, Roger and Edward Wakeling (eds.), *Lewis Carroll Photographer: The Princeton*

University Library Albums, Princeton, NJ: Princeton University Press, 2002

THOMSON'S MIRROR GALVANOMETER

Finn, Bernard S., *Submarine Telegraphy: The Grand Victorian Technology*, London: HMSO, 1973

THE KEW PHOTOHELIOGRAPH

De La Rue, Warren, "On the Total Solar Eclipse of July 18th, 1860 Observed at Rivabellosa, near Miranda de Ebro, in Spain [Bakerian Lecture]," *Philosophical Transactions of the Royal Society*, Vol. 52 (1862): pp. 333–416

De La Rue, Warren, "Comparison of Mr. De La Rue's and Padre Secchi's Eclipse Photographs," *Proceedings of the Royal Society*, Vol. 13 (1864): pp. 442–44

King, Henry C., *The History of the Telescope*, London: Griffin, 1955; reprinted London: Dover, 1979

THE FIRST PLASTIC

Friedel, Robert D., *Men, Materials, and Ideas: A History of Celluloid*, Johns Hopkins University, PhD dissertation, 1977; Ann Arbor, MI: University Microfilms International, 1982

Katz, Sylvia, *Classic Plastics: From Bakelite to High-Tech: With a Collector's Guide*, London: Thames & Hudson and New York: Abrams, 1984

Kaufman, Morris, *The First Century of Plastics: Celluloid and Its Sequel*, London: Plastics Institute, 1963

Meikle, Jeffrey, *American Plastic: A Cultural History*, New Brunswick, NJ: Rutgers University Press, 1995

THE LENOIR GAS ENGINE

Cummins, C. Lyle, *Internal Fire: The Internal Combustion Engine 1673–1900*, Lake Oswego, OR: Carnot Press, 1976

Day, John, *Engines: The Search for Power*, London: Hamlyn and New York: St Martin's Press, 1980

Georgano, G.N., *Cars: Early and Vintage 1886–1930*, London: Grange Universal, 1990

Robinson, William, *Gas & Petroleum Engines: A Practical Treatise on the Internal Combustion Engine*, New York and London: Spon, 1890; 2nd edition, 1902

Wise, David Burgess, "Lenoir: The Motoring Pioneer," in *The World of Automobiles*, London: Orbis, 1974

BESSEMER CONVERTER

Barraclough, Kenneth C., *Steelmaking 1850–1900*, London: Institute of Metals, 1990

Bessemer, Henry, *An Autobiography*, London: Offices of Engineering, 1905; Facsimile edition, London and Brookfield, VT: Institute of Metals, 1989

Gale, W.K.V., *Iron and Steel*, London: Longman, 1969

Moore, C. and R.I. Marshall, *Steelmaking*, London and Brookfield, VT: Institute of Metals, 1991

SWAN AND RAVEN

Wright, Thomas, "Scale Models, Similitude, and Dimensions: Aspects of Mid-Nineteenth-Century Engineering Science," *Annals of Science*, Vol. 49, No. 3 (1992): pp. 233–54

HOLMES' LIGHTHOUSE GENERATOR

Douglass, J.N., "The Electric Light Applied to Lighthouse Illumination," *Minutes of the Proceedings of the Institution of Civil Engineers*, Vol. 57 (1878): pp. 77–110

King, W.J., "The Development of Electrical Technology in the 19th Century: 3. The Early Arc Light and Generator," Contributions from the Museum of History and Technology, Washington, DC: *United States National Museum Bulletin*, Smithsonian Institution, No. 228, Paper 30 (1962): pp. 333–407

JULIA MARGARET CAMERON'S *IAGO*

Cox, Julian and Colin Ford, *Julia Margaret Cameron: The Complete Photographs*, Los Angeles: Getty Publications, 2003

Ford, Colin, *The Cameron Collection*, New York: Van Nostrand Reinhold and London: National Portrait Gallery, 1975

Howarth, Sophie and M. Darsie Alexander, *Singular Images: Essays on Remarkable Photographs*, New York: Aperture, 2005

Lukitsh, Joanne, *Julia Margaret Cameron*, London: Phaidon, 2006

Olsen, Victoria C., *From Life: Julia Margaret Cameron and Victorian Photography*, Basingstoke, Hampshire: Palgrave Macmillan, 2003

CROOKES' RADIOMETER

Crookes, William, "Molecular Physics in High Vacua," *Proceedings of the Royal Institution of Great Britain*, Vol. 9 (1879–81): pp. 138–59

Crookes, William, "On Radiant Matter," *Nature*, Vol. 20 (1879): pp. 419–23, 436–40

Crookes, William, "On the Illumination of Lines of Molecular Pressure and the Trajectory of Molecules," *Philosophical Transactions of the Royal Society*, Vol. 170 (1879): pp. 135–64

Fournier D'Albe, E.E., *The Life of Sir William Crookes*, London: Unwin, 1923

BELL'S OSBORNE TELEPHONE

Bruce, Robert V., *Bell: Alexander Graham Bell and the Conquest of Solitude*, London: Gollancz and Boston: Little Brown, 1973

Pool, Ithiel de Sola (ed.), *The Social Impact of the Telephone*, Cambridge, MA: MIT Press, 1977

WIMSHURST'S ELECTROSTATIC MACHINE

Marshall, Alfred W., *The Wimshurst Machine: How to Make It and Use It: A Practical Handbook on the Construction and Working of the Wimshurst Machine*, New York: Spon, 1908; facsimile edition, Bradley, IL: Lindsay Publications, 1989

Rowbottom, Margaret and Charles Susskind, *Electricity and Medicine: History of their Interaction*, London: Macmillan and San Francisco: San Francisco Press, 1984

THE FIRST TURBO-GENERATOR

Bowers, Brian P., *A History of Electric Light and Power*, New York: Peter Peregrinus (in association with the Science Museum, London), 1982

Scaife, W. Garret, "The Parsons Steam Turbine," *Scientific American*, Vol. 252, No. 4 (April 1985)

ROVER SAFETY BICYCLE

Bartleet, H.W., *Bartleet's Bicycle Book*, London: Burrow & Co., 1928; reprinted, Birmingham: John Pinkerton, 1983

Caunter, C.F., *The History and Development of Cycles. Part I*, London: HMSO, 1955

Caunter, C.F., *Handbook of the Collection Illustrating Cycles. Part II*, London: HMSO, 1958

The Cyclist and Wheel World Annual, 1862

Sharp, Archibald , *Bicycles & Tricycles: An Elementary Treatise on Their Design and Construction*, London: Longman, 1896; Cambridge, MA: MIT Press, 1977

Starley, K.J., "The Evolution of the Cycle," *Journal, Society of Arts*, Vol. 46, No. 2374 (May 20, 1898)

Sturmey, Henry, *Sturmey's Indispensable Handbook to the Safety Bicycle, Treating of Safety Bicycles, Their Varieties, Construction & Use*, Iliffe & Son, 1885; reprinted Birmingham: John Pinkerton, 1982

THE KODAK CAMERA

Coe, Brian, *Kodak Cameras: The First Hundred Years*, Hove: Hove Foto Books, 1988

Collins, Douglas, *The Story of Kodak*, New York: Abrams, 1990

Ford, Colin (ed.), *The Kodak Museum: The Story of Popular Photography*, London: Century, 1989

EARLY CINE-CAMERAS

Rawlence, Christopher, *The Missing Reel: The Untold Story of the Lost Inventor of Moving Pictures*, London: Collins, 1990

LOCOMOTIVE FROM THE FIRST LONDON "TUBE" RAILWAY

Holman, Printz P., *The Amazing Electric Tube: A History of the City and South London Railway*, London: Transport Museum, 1990

Horne, M.A.C., *The Northern Line: A Short History*, London: Douglas Rose/Nebulous Books, 1987

Jackson, Alan Arthur and Desmond F. Croome, *Rails Through the Clay: A History of London's Tube Stations*, London: Allen & Unwin, 1962; reprinted Harrow Weald, Greater London: Capital Transport, 1993

Lascelles, T.S., *The City & South London Railway*, Lingfield, Surrey: The Oakwood Press, 1955

PARSONS' MARINE STEAM TURBINE

Richardson, Alexander, *The Evolution of Parsons' Steam Turbine*, London: Offices of Engineering, 1911

PANHARD ET LEVASSOR CAR

The Saturday Review, July 20, 1895: pp. 79–80

COMING SOUTH, PERTH STATION

Daniels, Stephen, *Train Spotting: Images of the Railway in Art*, Nottingham: Nottingham Castle Museum, 1985

Ellis, C. Hamilton, *Railway Art*, edited by Susan Hyman, Boston: Ash & Grant and New York: Graphic Society, 1977

Noakes, Aubrey, *William Frith: Extraordinary Victorian Painter*, London: Jupiter, 1978

Richards, Jeffrey and John M. Mackenzie, *The Railway Station: A Social History*, Oxford and New York: Oxford University Press, 1986

THOMSON'S CATHODE-RAY TUBE

Hey, Anthony J.G. and Patrick Walters, *The Quantum Universe*, Cambridge and New York: Cambridge University Press, 1987

Pais, Abraham, *Inward Bound*, Oxford: Clarendon Press and New York: Oxford University Press, 1986

MARCONI'S FIRST TUNED TRANSMITTER

Baker, W.J., *A History of the Marconi Company 1874-1965*, London: Routledge, 1970

Eastwood, Eric (ed.), *Wireless Telegraphy* (Royal Institution Library of Science), London: Applied Science, 1974

POULSEN'S TELEGRAPHONE

Camras, Marvin (ed.), *Magnetic Tape Recording*, New York: Van Nostrand Rheinhold, 1985

"The Telegraphone," *The Electrician*, April 26, 1901: pp. 5–7 and July 31, 1903: pp. 611–12

FLEMING'S ORIGINAL THERMIONIC VALVES

Fleming, J.A., "A Further Examination of the Edison Effect in Glow Lamps," *Proceedings of the Physical Society of London*, Vol. 14, Issue 1, pp. 187–242 (1895)

MASS PRODUCTION AND THE MODEL T FORD

Casey, Robert, *The Model T: A Centennial History*, Baltimore: Johns Hopkins University Press, 2008

Clark, William, Jan Golinski and Simon Schaffer (eds.), *The Sciences in Enlightened Europe*, Chicago: University of Chicago Press, 1999

Hall, Karyl Lee Kibler and Carolyn C. Cooper, *Windows on the Works: Industry on the Eli Whitney Site 1798–1979*, Hamden, CT: Eli Whitney Museum, 1984

Hooker, Clarence, *Life in the Shadow of the Crystal Palace 1910–1927: Ford Workers in the Model T Era*, Bowling Green, OH: Bowling Green State University Popular Press, 1997

Hounshell, David A., *From the American System to Mass Production 1800–1932: The Development of Manufacturing Technology in the United States*, Baltimore: Johns Hopkins University Press, 1984

Lacey, Robert, *Ford: The Men and the Machine*, Boston: Little Brown, 1986

McCalley, Bruce W., *Model T Ford: The Car that Changed the World*, Iola, WI: Krause, 1994

Rider, Christine (ed.), *Encyclopedia of the Age of the Industrial Revolution*, 2 volumes, Westport, CT: Greenwood Press, 2007

Smil, Vaclav, *Creating the Twentieth Century: Technical Innovations of 1867–1914 and Their Lasting Impact*, Oxford and New York: Oxford University Press, 2005

HABER'S SYNTHETIC AMMONIA

Haber, L.F., *The Chemical Industry 1900–1930: International Growth and Technological Change*, Oxford: Clarendon Press, 1971

THE MUNITIONS GIRLS

Baker, Henry, *The "Steel" Bakers of Rotherham: Being a Brief History of the Baker Family and the Business of John Baker & Bessemer Ltd.*, Rotherham, 1960; revised edition, 2004

Wilson, R.T. and E.J. Twigg, *Industrial Sheffield and Rotherham: The Official Handbook of the Sheffield and Rotherham Chambers of Commerce*, Derby: Bemrose, 1919

ASTON'S MASS SPECTROGRAPH

Aston, F.W., "A Simple Form of Micro-Balance for Determining the Densities of Small Quantities of Gases," *Proceedings of the Royal Society of London*, Series A, Vol. 89 (1914): pp. 439–46

Gillispie, Charles Coultston (ed.), *Dictionary of Scientific Biography*, Vol. 1, New York: Scribner, 1970, pp. 320–22

Magill, Frank N. (ed.), *The Nobel Prize Winners: Chemistry*, Pasadena, CA: Salem Press, 1990, pp. 250–57

ALCOCK AND BROWN'S VICKERS VIMY

Alcock, John and Arthur Whitten Brown, *Our Transatlantic Flight*, London: Kimber, 1969

AUSTIN SEVEN PROTOTYPE CAR

Wyatt, Robert John, *The Austin Seven: The Motor for the Million 1922–1939*, Newton Abbot, Devon: David & Charles, 1968; 3rd edition, 1982

DOBSON SPECTROPHOTOMETER

Dobson, G.M.B., "Forty Years' Research on Atmospheric Ozone at Oxford: A History," *Applied Optics*, Vol. 7, Issue 3, pp. 387–405 (March 1968)

LOGIE BAIRD'S TELEVISION APPARATUS

Burns, R.W., *British Television: The Formative Years*, London: Peter Peregrinus (in association with the Science Museum, London), 1986

Burns, R.W., *John Logie Baird: Television Pioneer*, London: Institution of Electrical Engineers, 2000

Garratt, G.R.M., *Television: An Account of the Development and General Principles of Television as Illustrated by a Special Exhibition Held at the Science Museum*, London: Science Museum, 1937

PENICILLIN

Brown, Kevin, *Penicillin Man: Alexander Fleming and the Antibiotic Revolution*, Stroud, Gloucestershire: Sutton, 2004

Fleming, Alexander, "On the Antibacterial Action of Cultures of a Penicillium, with Special Reference to Their Use in the Isolation of *B. Influenzæ*," *British Journal of Experimental Pathology*, Vol. 10, No. 3 (1929): pp. 226–36

Hare, Ronald, *The Birth of Penicillin and the Disarming of Microbes*, London: Allen & Unwin, 1976

Maurois, André, *The Life of Sir Alexander Fleming, Discoverer of Penicillin*, New York: Dutton, 1959

SUPERMARINE S.6B FLOATPLANE

Grey, C.G. and Leonard Bridgman (eds.), *Jane's All the World's Aircraft*, London: Sampson Low, Marston, 1931

Orlebar, Augustus Henry, *Schneider Trophy*, London: Seeley, 1933

LAWRENCE'S ELEVEN-INCH CYCLOTRON

Close, Frank E., Michael Marten and Christine Sutton, *The Particle Explosion*, Oxford and New York: Oxford University Press, 1987

Rhodes, R., *The Making of the Atomic Bomb*, New York: Simon and Schuster, 1986

COCKCROFT AND WALTON'S ACCELERATOR

Crowther, J.G., *The Cavendish Laboratory 1874–1974*, London: Macmillan and New York: Science History Publications, 1974

CHADWICK'S PARAFFIN WAX

Segrè, Emilio, *From X–rays to Quarks: Modem Physicists and Their Discoveries*, San Francisco: W.H. Freeman, 1980

THE DISCOVERY OF POLYETHYLENE

Kennedy, Carol, *ICI: The Company that Changed Our Lives*, London: Hutchinson, 1986; 2nd edition, London: P. Chapman, 1993

THE BOEING 247D

Berry, Peter, *The Douglas Commercial Story*, Saffron Walden, Essex: Air-Britain Publications, 1971

Taylor, John William Ransom and Kenneth Munson, *Air Transport Before the Second World War*, London: New English Library, 1975

DISCOVERY OF ARTIFICIAL RADIOACTIVITY

Weart, Spencer R., *Scientists in Power*, Cambridge, MA: Harvard University Press, 1979

REYNOLDS' X-RAY MACHINE

Reynolds, Russell J., "Cineradiography," *Institution of Electrical Engineers Journal*, Vol. 79, No. 478 (October 1936): pp. 389–400

ORIGINAL RADAR RECEIVER

Bowen, E.G., *Radar Days*, Bristol: Adam Hilger, 1987

Burns, R.W. (ed.), *Radar Development to 1945*, London: Peter Peregrinus (in association with the Institution of Electrical Engineers), 1988

Watson-Watt, Robert, *Three Steps to Victory: A Personal Account by Radar's Greatest Pioneer*, London: Odhams Press, 1957

MANCHESTER DIFFERENTIAL ANALYZER

Anon., "Machine Solves Mathematical Problems: A Wonderful Meccano Mechanism," *The Meccano Magazine*, No. 19 (June 1934): pp. 442–44

Augarten, Stan, *Bit by Bit: An Illustrated History of Computers*, London: Allen & Unwin and New York: Ticknor & Fields, 1984

Bush, Vannevar, *Pieces of the Action*, New York: Morrow, 1970, p. 161

Crank, John, *The Differential Analyser*, London: Longman, 1947

Hartree, D.R., "A Great Calculating Machine: The Bush Differential Analyser and Its Applications in Science and Industry," *Proceedings of the Royal Institution of Great Britain*, Vol. 31, (1939–40): p. 151

Hartree, D.R., "The Differential Analyser," *Nature*, Vol. 135 (June 1935): pp. 940–43

EMITRON CAMERA TUBE

Burns, R.W., *British Television: The Formative Years*, London: Peter Peregrinus (in association with the Science Museum, London), 1986

Norman, Bruce, *Here's Looking At You: The Story of British Television 1908–1939*, London: Royal Television Society/BBC, 1984

Pawley, Edward, *BBC Engineering 1922–1972*, London: BBC, 1972

MALLARD

Allen, C.J., *The Gresley Pacifics of the LNER*, London: Ian Allan, 1950

Bellwood, John and David Jenkinson, *Gresley and Stanier*, London: HMSO, 1976; 2nd edition, 1986

Nock, O.S., *The Gresley Pacifics*, Newton Abbot, Devon: David & Charles, 1982

Rutherford, Michael, *Mallard the Record Breaker*, York: Newburn House/Friends of the National Railway Museum, 1988; 2nd edition, 2004

THE BAKELITE COFFIN

Ellis, William R., "Material of a Thousand Uses," *Ohio State Engineer*, Vol. 17, No. 5 (March 1934): pp. 3–4

Katz, Sylvia, *Classic Plastics: From Bakelite to High-Tech: With a Collector's Guide*, London: Thames & Hudson and New York: Abrams, 1984

Mumford, John Kimberly, *The Story of Bakelite*, New York: Robert L. Stillson, 1924

RANDALL AND BOOT'S CAVITY MAGNETRON

Rolph, P.M. (ed.), *Fifty Years of the Cavity Magnetron: Proceedings of a One-Day Symposium 21 February 1990*, School of Physics and Space Research, Birmingham: University of Birmingham, 1991

Wathen, Robert L., "Genesis of a Generator: The Early History of the Magnetron," *Journal of the Franklin Institute*, Vol. 255, No. 4 (April 1953): pp. 271–87

THE ROLLS-ROYCE MERLIN

Harvey-Bailey, Alec and Dave Piggott, *The Merlin 100 Series: The Ultimate Military Development*, Derby: Rolls-Royce Heritage Trust, 1993

Harvey-Bailey, Alec, *The Merlin in Perspective: The Combat Years*, Derby: Rolls-Royce Heritage Trust, 1983

Pugh, Peter, *The Magic of a Name: The Rolls-Royce Story, The First 40 Years*, Cambridge: Icon Books and New York: Totem Books, 2000

Setright, L.J.K., *The Power to Fly: The Development of the Piston Engine in Aviation*, London: Allen & Unwin, 1971

GLOSTER-WHITTLE E.28/39 JET AIRCRAFT

Golley, John, *Whittle: The True Story*, Airlife, 1987; as *Genesis of the Jet: Frank Whittle and the Invention of the Jet Engine*, Ramsbury, Marlborough, Wiltshire: Crowood Press, 1996

Naham, Andrew, *Frank Whittle: Invention of the Jet*, Cambridge: Icon Books, 2004

Whittle, Frank, *Jet: The Story of a Pioneer*, London: Frederick Muller, 1953

THE V-2 ROCKET

Dungan, T.D., *V-2: A Combat History of the First Ballistic Missile*, Yardley, PA: Westholme, 2005

Kennedy, Gregory P., *Germany's V-2 Rocket*, Atglen, PA: Schiffer Books, 2006

Kennedy, Gregory P., *Vengeance Weapon 2: The V2 Guided Missile*, Washington, DC: National Air and Space Museum/Smithsonian Institution Press and London: Harper Collins, 1983

Ordway III, Frederick and Mitchell R. Sharpe, *The Rocket Team*, London: Heinemann and New York: Crowell, 1979

THE FIRST MARINE GAS TURBINE

Kerrebrock, Jack L., *Aircraft Engines and Gas Turbines*, Cambridge, MA: MIT Press, 1992

Phelan, Keiran and Martin Hubert Brice, *Fast Attack Craft: The Evolution of Design and Tactics*, London: Macdonald and Jane's, 1977

VITAMIN B12 MODEL

Dowd, Paul, "On the Mechanism of Action of Vitamin B12. Model Studies. Thermal Rearrangement of Methyl 3, 3-Dimethylglycidate to Methyl Levulinate," *Proceedings of the National Academy of Sciences of the United States of America*," Vol. 69, No. 5 (May 1972): pp. 1173–75

Pratt, J.M., *Inorganic Chemistry of Vitamin B12*, New York: Academic Press, 1972

Schrauzer, G.N., *Bioinorganic Chemistry, Vitamin B 12 and Vitamin B 12 Model Compounds*, Washington, DC: American Chemistry Society, 1970

Smith, E. Lester, *Vitamin B 12*, London: Methuen and New York: John Wiley, 1965

Stagni, N., B. de Bernard, G. Costa and G. Mestroni, "Biological Properties of Model Molecules for Vitamin B12," *Nature*, Vol. 225 (March 7, 1970): pp. 942–43

Woodward, Robert Burns, Boleslaw Zagalak, "Vitamin B12," *Proceedings of the 3. European Symposium on Vitamin B12 and Intrinsic Factor*, University of Zurich (March 1979)

PILOT AUTOMATIC COMPUTING ENGINE

Campbell-Kelly, Martin, "Programming the Pilot ACE: Early Programming Activity at the National Physical Laboratory," *Annals of the History of Computing*, Vol. 3, No. 2 (April–June 1981): pp. 133–62

Campbell-Kelly, Martin and William Aspray, *Computer: A History of the Information Machine*, New York: Basic Books, 1996

The Early British Computer Conferences, Vol. 14, Charles Babbage Institute reprint series, Cambridge, MA: MIT Press, 1989

Williams, Michael R., *A History of Computing Technology*, Englewood Cliffs, NJ: Prentice Hall, 1985

FESTIVAL OF BRITAIN PATTERNS

Festival Pattern Group, The, *Souvenir Book of Crystal Designs: The Fascinating Story in Colour of the Festival Pattern Group*, London: HMSO, 1951

Forgan, Sophie, "Festivals of Science and the Two Cultures: Science, Design and Display in the Festival of Britain 1951," *British Journal for the History of Science*, Vol. 31 (1998): pp. 217–40

Jackson, Lesley, *From Atoms to Patterns: Crystal Structure Designs from the 1951 Festival of Britain*, London: Richard Dennis/Wellcome Trust, 2008

Megaw, Helen, *Crystal Structures: A Working Approach*, Philadelphia and London: Saunders, 1973

NUCLEAR MAGNETIC RESONANCE

Akitt, J.W. and B.E. Mann, *NMR and Chemistry: An Introduction to Modern NMR Spectroscopy*, London: Chapman and Hall, 1973; 4th edition, Cheltenham, Gloucestershire: Stanley Thornes, 2000

Grant, David M. and Robin K. Harris (eds.), *Encyclopedia of Nuclear Magnetic Resonance*, 9 volumes, Chichester, West Sussex and New York: John Wiley, 2002

Keeler, James, *Understanding NMR Spectroscopy*, Chichester, West Sussex and New York: John Wiley, 2005

Sutton, Christine, "A Magnetic Window into Bodily Functions," *New Scientist*, Vol. 3, No. 1525 (September 1986): pp. 32–37

CRICK AND WATSON'S DNA MODEL

Olby, Robert, *The Path to the Double Helix*, London: Macmillan and Seattle: University of Washington Press, 1974

Watson, James D. and F.H.C. Crick, "Molecular Structure of Nucleic Acids: A Structure for Deoxyribose Nucleic Acid," *Nature*, Vol. 171 (April 1953): pp. 737–38

Watson, James D., *The Double Helix: A Personal Account of the Discovery of the Structure of DNA*, London: Weidenfeld & Nicolson and New York: Atheneum, 1968; reprinted with additional articles, edited by Gunther S. Stent, New York: Norton, 1980 and London: Weidenfeld & Nicolson, 1981; reprinted, New York: Scribner, 1998; reprinted, with an introduction by Sylvia Nasar, New York: Touchstone, 2001

FLYING BEDSTEAD TEST RIG

Rogers, Mike, *VTOL Military Research Aircraft*, Yeovil, Somerset: Haynes Foulis Aviation and New York: Orion Books, 1989

THE CESIUM ATOMIC CLOCK

Audoin, Claude and Bernard Guinot, *The Measurement of Time: Time, Frequency and the Atomic Clock*, Cambridge: Cambridge University Press, 2001

Essen, L. and J.V.L. Parry, "The Caesium Resonator as a Standard of Frequency and Time," *Philosophical Transactions of the Royal Society*, Vol. 250 (1957): pp. 45–69

Forman, P., "Atomichron: The Atomic Clock from Concept to Commercial Product," *Proceedings of the IEEE*, Vol. 73, No. 7 (1985): pp. 1181–1204

THE WORLD'S FIRST HOVERCRAFT

Crewe, P.R., "The Hovercraft: A New Concept in Maritime History," *Transactions of the Royal Institution of Naval Architects*, Vol. 102 (1960): pp. 315–65

AMPEX VR-1000 VIDEO RECORDER

Pawley, Edward, *BBC Engineering 1922–1972*, London: BBC, 1972

Robinson, Joseph F., *Videotape Recording: Theory and Practice*, New York: Focal Press and London: Hastings House, 1975

FERRANTI PEGASUS COMPUTER

Lavington, Simon, *Early British Computers: The Story of Vintage Computers and the People Who Built Them*, Manchester: Manchester University Press and Bedford, MA: Digital Press, 1980

Lavington, Simon, *The Pegasus Story: A History of a Vintage British Computer*, London: Science Museum, 2000

ELECTRON-CAPTURE DETECTOR

Chen, E.C.M. and E.S.D. Chen, *The Electron Capture Detector and the Study of Reactions with Thermal Electrons*, Hoboken, NJ: John Wiley, 2004

Lovelock, J.E., "The Electron Capture Detector," *Journal of Chromatography*, Vol. 99, No. 1 (1974): pp. 3–12

Pellizzari, E.D., "Electron Capture Detection in Gas Chromatography," *Journal of Chromatography*, Vol. 98, No. 2 (1974): pp. 323–36

INTERNAL CARDIAC PACEMAKER

Greatbatch, Wilson, *The Making of the Pacemaker: Celebrating a Lifesaving Invention*, Amherst, NY: Prometheus Books, 2000

Portal, R.W., J.G. Davies, A. Leatham, A.H.M. Siddons, "Artificial Pacing for Heart Block," *The Lancet*, Vol. 2 (1962): pp. 1369–75

SCANNING ELECTRON MICROSCOPE

Oatley, C.W., "The Early History of the Scanning Electron Microscope," *Journal of Applied Physics*, Vol. 53, Issue 2 (1982): pp. R1–13

Stewart, A.D.G., "The Origins and Development of Scanning Electron Microscopy," *Journal of Microscopy*, Vol. 139 (1985): pp. 121–27

ULTRASOUND SCANNER

Blume, Stuart S., *Insight and Industry: On the Dynamics of Technological Change in Medicine*, Cambridge, MA: MIT Press, 1992

Oakley, Ann, *The Captured Womb: A History of the Medical Care of Pregnant Women*, Oxford and New York: Basil Blackwell, 1984

Yoxen, Edward, "Seeing with Sound: A Study of the Development of Medical Images," *The Social Construction of Technological Systems: New Directions in the Sociology and History of Technology*, edited by Wiebe E. Bijker *et al*, Cambridge, MA: MIT Press, 1987

THE MOLINS SYSTEM 24: THE FIRST FLEXIBLE MANUFACTURING SYSTEM

Talavage, Joseph and Roger G. Hannam, *Flexible Manufacturing Systems in Practice: Applications, Design and Simulation*, Boca Raton, FL: CRC Press, 1988

Williamson, D.T.N., "Molins System 24—A New Concept of Manufacture," *Machinery and Production Engineering*, September 13, 1967: pp. 544–55 and October 25, 1967: pp. 852–63

APOLLO 10 COMMAND MODULE

Chaikin, Andrew, *A Man on the Moon: The Voyages of the Apollo Astronauts*, New York and London: Viking, 1994

Godwin, Robert (ed.), *Apollo 10: The NASA Mission Reports*, 2nd edition, Burlington, Ontario: Apogee Books, 2000

Lattimer, Dick, *All We Did Was Fly to the Moon*, Gainesville, FL: Whispering Eagle Press, 1985

CONCORDE 002

Edgerton, David, *England and the Aeroplane: An Essay on a Militant and Technological Nation*, London: Macmillan, 1991

Kelly, Neil, *The Concorde Story: 34 Years of Supersonic Air Travel*, West Molesey, Surrey: Merchant Book Co., 2005

March, Peter R., *The Concorde Story*, Stroud, Gloucestershire: History Press, 2005

Orlebar, Christopher, *The Concorde Story*, 7th edition, London: Osprey, 2011

Owen, Kenneth, *Concorde: New Shape in the Sky*, London: Jane's/Science Museum, 1982

ROLLS-ROYCE RB211

Gunston, Bill, *Development of Piston Aero Engines*, Wellingborough, Northamptonshire: Patrick Stephens, 1993; 2nd edition, 2006

Gunston, Bill, *Rolls-Royce Aero Engines*, Wellingborough, Northamptonshire: Patrick Stephens, 1989

Hooker, Stanley, *Not Much of an Engineer: An Autobiography*, Shrewsbury: Airlife and Warrendale, PA: Society of Automotive Engineers, 1984

THE FIRST BRAIN SCANNER

Beckmann, E.C., "CT Scanning: The Early Days," *The British Journal of Radiology*, Vol. 79, No. 937 (January 2006): pp. 5–8

Blume, Stuart S., *Insight and Industry: On the Dynamics of Technological Change in Medicine*, Cambridge, MA: MIT Press, 1992

Gunderman, Richard B., *Essential Radiology: Clinical Presentation, Pathophysiology, Imaging*, New York: Thieme, 2006

Hounsfield, G.N., "Computerized Transverse Axial Scanning (Tomography): Part I. Description of System," *British Journal of Radiology*, Vol. 46 (1973): pp. 1016–22

Susskind, Charles, "The Invention of Computed Tomography," in *History of Technology*, edited by A. Rupert Hall and Norman Smith, Vol. 6, London: Mansell, 1981, pp. 39–80

MAGNETIC RESONANCE IMAGING

Brown, Mark A. and Richard A. Semelka, *MRI*, Hoboken, NJ: Wiley-Blackwell, 2010

Hashemi, Ray H., William G. Bradley, Jr. and Christopher J. Lisanti, *MRI: The Basics*, Philadelphia: Lippincott, Williams & Wilkins, 2010

Lees, S.C. *et al.*, "One Micrometer Resolution NMR Microscopy," *Journal of Microscopy*, Vol. 150, Issue 2 (June 2001): pp. 207–13

Mattson, James and Merrill Simon, *The Pioneers of NMR and Magnetic Resonance in Medicine: The Story of MRI*, Ramat Gan, Israel: Bar-Ilan University Press and Westhampton, NY: Dean Books, 1996

DAVID HOCKNEY: *NATIONAL MEDIA MUSEUM PHOTO COLLAGE*

Gayford, Martin, *A Bigger Message: Conversations with David Hockney*, New York: Thames & Hudson, 2011

Hockney, David and Paul Joyce, *Hockney on Photography: Conversations with Paul Joyce*, New York: Harmony, 1988

GENETICALLY ENGINEERED MICE

Harris, John (ed.), *Bioethics*, New York and Oxford: Oxford University Press, 2001

Krimsky, Sheldon, *Biotechnics and Society: The Rise of Industrial Genetics*, New York: Praeger, 1991

Mills, Oliver, *Biotechnical Inventions: Moral Restraints and Patent Law*, Farnham, Surrey: Ashgate, 2010

Sundberg, John P. and Tsutomu Ichiki (eds.), *Genetically Engineered Mice Handbook*, Boca Raton, FL: CRC Press and London: Taylor & Francis, 2006

PROTOTYPE "CLOCK OF THE LONG NOW"

Brand, Stewart, *The Clock of the Long Now: Time and Responsibility: The Ideas Behind the World's Slowest Computer*, New York: Basic Books, 1999

Rose, Alexander, *The Clock of the Long Now: Mechanical Drawings of the First Prototype*, Seattle: CreateSpace, 2010

INDEX

electrons, 28, 110, 132, 138, 162, 164, 170, 184, 214, 218

electrophorus, 114

electrostatic machine, Wimshurst's, 114-15

electrotherapy, 114

Eli Lilly and Company, 194

Elizabeth I, queen of England, 22

Elizabeth II, queen of Great Britain and Northern Ireland, 128, 238

Elliott Brothers (London) Ltd., 212

Ellis, Alexander John, 82

Ellis, Evelyn Henry, 128

Elmqvist, Rune, 216

Elton, Michael, 80

EMI, *see* Electric and Musical Industries Ltd.

EMIDEC 1100, 230

Emitron camera tube, 154, 178-79

energy (measurement of), 88, 110; *see also* joule

Engineer, The (journal), 100

engineering, 24, 74, 116, 124, 184, 186, 192, 228

engine(s) (aircraft):
    Alvis Leonides, 208
    Avon turbojet (Rolls-Royce), 226
    Eagle (Rolls-Royce), 186
    Eagle VIII (Rolls-Royce), 148
    F2/3 (Metropolitan-Vickers), 192
    gas turbine, 188, 204, 228
    jet, 158, 188, 192, 204, 208, 226, 228-29
    Kestrel (Rolls-Royce), 186
    Merlin (Rolls-Royce), 158, 186-87, 228
    Napier Lion, 158
    Nene (Rolls-Royce), 204
    Olympus 593 Mk 3B (Rolls-Royce), 228
    "R" (Rolls-Royce), 158, 186
    RB108 (Rolls-Royce), 204
    RB211 (Rolls-Royce), 228-29
    W.1 (Power Jets), 188

engine(s) (non-aircraft): *see also* computing; machinery
    Comet steamship (Bell/Robertson), 44, 60-61
    dividing (Troughton), 38-39

first turbo-generator (Parsons), 116-17, 126

G1 marine gas turbine (Metropolitan-Vickers), 192

G2 marine gas turbine (Metropolitan-Vickers), 192

gas (Lenoir), 100-101

high-pressure (Trevithick), 56-57, 60, 62

marine gas turbine, first, 192-93

marine steam turbine (Parsons), 126-27, 192

marine (Symington), 44-45

Newcomen, 14, 42, 44, 56

rotative (Boulton and Watt), 10, 42-43, 56, 60

English Mechanic and World of Science, The (magazine), 172

"Enlightened Automata" (essay by Schaffer from *The Sciences in Enlightened Europe*), 140

Enlightenment, the, 140

Eno, Brian, 238

Erector (model construction set), 176

Eschenmoser, Albert, 194

Essen, Louis, 206

European Organization for Nuclear Research (CERN), 162

European Patent Office, 236

Ewart, Peter, 58

Excursions Daguerriennes (work by Lerebours), 82

F2/3 aircraft engine, 192

Fabry, Maurice Paul Auguste Charles, 152

factory system, 36, 42, 54, 140, 222; *see also* manufacturing; mass production

Fairey Delta 2 (aircraft), 226

Faraday, Michael, 66, 106

Faraday, Robert, 66

Farington, Joseph, 52

Farman, Joseph C., 152

farming, *see* agriculture

Fawcett, Eric, 166

Felling Colliery, 64

Ferguson, Adam, 140

Fermi, Enrico, 164, 170

Ferranti Ltd., 212, 222

fertilizer, 142

Festival of Britain (1951), 12, 172, 198

Festival of Britain patterns, 198-99

fetus measurement (in ultrasound), 220

Field, Mary Katherine Keemle, 112

First Aero Engine Made by Rolls-Royce Limited, The (work by Royce), 186

Flamsteed, John, 40

Fleming, Sir Alexander, 156

Fleming, Sir John Ambrose, 138

flexible manufacturing systems, 222-23

Fliegende Hamburger (locomotive), 180

Florey, Sir Howard Walter, 156

Flying Bedstead test rig, 204, 228

Flying Scotsman (locomotive), 180

FMS, *see* flexible manufacturing systems

Folkers, Karl August, 194

For King and Country (artwork by Skinner), 144

Forbes, Stanhope Alexander, 144

Ford, Henry, 140

Ford Motor Company, 140, 186

Forging the Anchor (artwork by Forbes), 144

Forman, Paul, 206

Forth & Clyde Canal Navigation Co., 44

Foulis, Sir William, 80

Fourier Transform spectrometer, 200

Frankland, Sir Edward, 146

Franklin, Rosalind Elsie, 202

French Revolution, 54

friction, 104, 114, 116, 208, 226

Frith, William Powell, 130

From the American System to Mass Production 1800-1932 (work by Hounshell), 140

Froude, Edmund, 104

Froude, William, 104, 192

Froude's hypothesis, 104

"Future Developments in Aircraft Design" (thesis by Whittle), 188

G1 marine gas turbine, 192

G2 marine gas turbine, 192